Distortion

The Cause Of Harmonics And The Lie Of THD

By Dan P. Bullard

A book that explains the cause of harmonic distortion in electronic circuits.

First Edition

Disclaimer

The author has made every effort to verify the information contained herein; however, errors may exist. Neither the author nor the publisher, CreateSpace, are responsible for damages arising from use of this information.

Cover art:
A montage of waveforms and spectra from the book distorted with a Fish Eye effect.

Other books by Dan P. Bullard

The Reluctant Road Warrior
The adventures of a Silicon Valley engineer in the U.S. and abroad

Harmonics
How they are made

Acknowledgements

The author would like to thank Skip Davis and Kars Schaapman of Applicos Measurement and Control for sponsoring the video and providing the computer that made this book possible. Thanks also to Bob Metzler of Audio Precision for the original inspiration for this project, and Don Greer of Maxim Integrated Products for setting the standard high enough to make me fully explain my findings. A special thank-you to Dr. John Michael Williams and Norman Abbod MSEE for proofreading. Thanks to David Reynolds who is always happy to chat with me on mixed signal and DSP topics and helped keep me on track. Thank you to Brent David Cartwright who painstakingly reviewed the book and asked lots of questions, which thankfully I had answers for. Also, to my lovely wife April for patiently listening to my theories and speculations, and putting up with my O-Dark-Thirty experiments.

My biggest thank-you goes out to Bruce Tibbetts of Teradyne, who inadvertently inspired me to dig so deeply into harmonics and their true cause.

For my wife, my children, and my grandchildren

False facts are highly injurious to the progress of science, for they often endure long

Charles Darwin, The Descent Of Man

Table of Contents

Introduction
Harmonics - A misunderstood phenomenon

Harmonics are the bane of many engineers. When stimulated by a sinusoid, a perfectly linear electronic circuit should output a sinusoid with no additional harmonic content. Everyone seems to know this, but the source of the harmonics in a non-perfect circuit seems to elude us. What causes the harmonics to form? What circuit faults cause which harmonics? Can I forgive a non-linearity knowing that the harmonics created will not impact my application? If I see harmonics can I get a clue to the problem I face from the *harmonic signature*?

Most texts on the subject make generalizations that are not very helpful. In Audio Precision's 1992 book *The Audio Measurement Handbook,* which I skewered in my YouTube video Audio Quality and Total Harmonic Distortion[1] the statements are made that:

"non-linearities which are not symmetrical around zero produce dominantly even

harmonics" and

"non-linearities which are symmetrical around zero produce odd harmonics"

Both of these statements are wrong (for different reasons), and my video proved this. Other books dealing with the topic are similarly flawed and vague, a bad combination for some poor soul working late into the night trying to figure out how to fix a high THD failure in his circuit.

This book will show you exactly how harmonics are created by non-linearities, which commonly found non-linearities create which harmonics, and how a harmonic signature can be interpreted to correctly assign blame to particular non-linearities allowing the engineer to spend his time wisely in identifying and fixing non-linearities in his circuit.

The information in this book was gleaned from experiments performed with an Excel spreadsheet that replicates the function of a sinusoid applied to a theoretical transfer function that can be manipulated at will, as well as years of working with real electronic circuits and various DSP based tools during my long career in electronics.

Throughout the book you will find links you can click on with your E-reader, which is one reason I published this as an E-book as well as a print book. If you are reading the print book, select hyperlinks are listed at the bottom of the page as footnotes (see below). Text is nice, but I have gone to a lot of trouble creating videos that can say more in a few seconds than hours of reading, so please take advantage of these links while reading this book. Throughout the book it is assumed that you have, at the very least, watched the video listed in the footnote below, for this video holds the key to this very interesting story of discovery—the story of how I figured out how harmonics are created. Join me now and prepare to be permanently changed by this journey.

---

[1] http://youtu.be/CHfeMGQC6Wl

A note on graphics

Virtually all of the figures in this book are one of two types. **Time domain** plots like the one below display waveforms, in which the X axis is in samples; usually 2048, and the Y axis is in volts; usually from -1V to +1V. The X axis labels are displayed so that the reader can compute angles or times to aid in comprehension of the material herein. The label

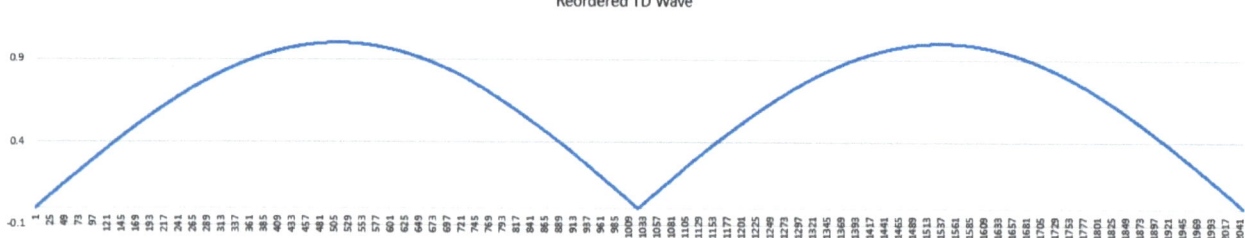

Reordered TD Wave refers to the fact that I reorder[2] 19 cycles into 1 cycle.

The other type of plot is a **frequency domain**, or spectral display, where the X axis is in spectral bin number and the Y axis is in dB, ranging from 0dB to -100 dB.

I used "Dan's Rules"[3] to defy Nyquist and "unfold" the first order aliasing harmonics to create a spectrum of 2048 bins from my 2048 time domain samples. The X axis labels are displayed to help the reader count harmonics if desired. The fundamental frequency is always in bin 19, so, for example, the 4th harmonic would be in bin 4*19, or bin 76. The highest harmonic displayed is the 107th in bin 2033. Bins are color coded **odd** and **even** and the legend noting this is always displayed.

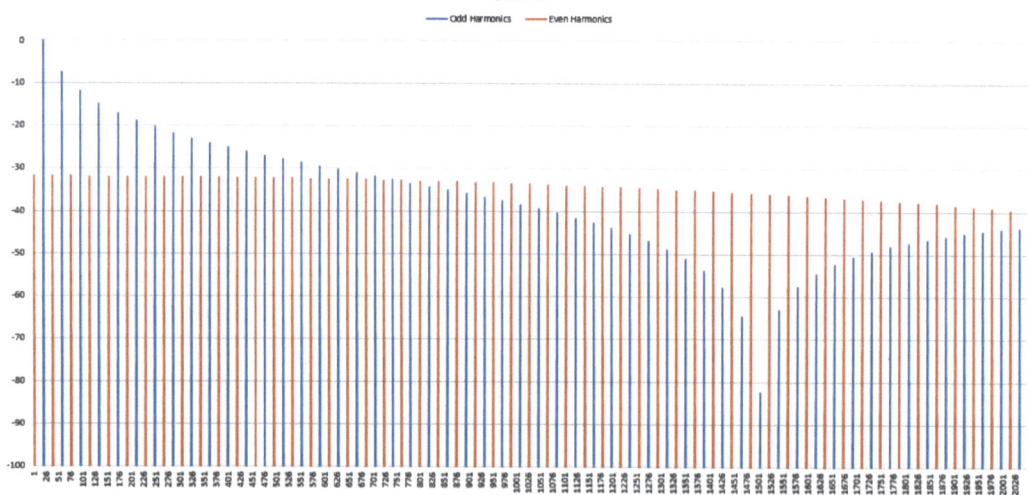

Both types of plots were created with Excel and are the result of purely mathematical waves being processed by Excel including the Fast Fourier Transform (FFT) from Microsoft's optional Data Analysis Pack.

---

[2] https://youtu.be/Gry57CQdik4

[3] https://youtu.be/vJ8V8ipSZ50

Chapter 1
The cause of harmonics - Area and Energy

When I posted the video <u>Audio Quality and Total Harmonic Distortion</u>[4] on YouTube and shared it on LinkedIn I was almost instantly hammered by an ex-coworker from my days at Credence Systems Corporation who had recently gotten an MSEE degree. My sponsor of the video, <u>Applicos Measurement and Control</u>[5] was a bit taken aback to say the least. Here was an experienced test engineer with a freshly minted MSEE degree challenging the video Applicos had paid me to make touting their product. This engineer insisted that we take the video down because I had misunderstood the statements in the Audio Precision book which therefore invalidated the thrust of the video.

Bruce Tibbetts
Applications Engineer at Teradyne

If I were you I would take that video down. Dan misunderstood what the author was saying and his demonstration is meaningless with regards to what was being discussed in the text.

Regards,
Bruce

Figure 1 - Linkedin post critical of my Audio Quality video

Shortly thereafter, a dialog began between Applicos and Bruce. Bruce claimed that harmonic content could not be changed by altering the DC offset of a sine wave (which proved he didn't get the point of the video), gave them numerous formulae that were intended to disprove my claims and asserted that I had no expertise on the subject. I was under the gun to prove that I was right. If I had made an error I would likely have to give back the money Applicos had paid me for the video and my utility as an expert in Mixed Signal testing would be nullified. I knew my statements were true, I had proven it time and time again in the past, but now I dove headlong into finding out *why* my statements in the video were true and why Audio Precision's statements were untrue.

---

[4] <u>http://youtu.be/CHfeMGQC6Wl</u>

[5] <u>http://www.applicos.com</u>

The key to understanding how harmonics are created by non-linearities came when I had to deal with the fact that non-linearities that are non-symmetrical around zero (volts, what AP meant in the book) created **even** harmonics (2nd, 4th, 6th, etc) in addition to the always present **odd** harmonics (1st, 3rd, 5th, 7th, etc). As mentioned in "A note on graphics" above, I will color code the words **odd** and **even** in this book to help you identify the harmonics in the graphs that follow.

You cannot have **even** harmonics without **odd** harmonics. The reason is simple. The fundamental is always the 1st harmonic, and as you might have figured out, the number 1 is **odd**!

There is one way to generate a waveform that consists of only **even** harmonics (sort of), and that is to perform full wave rectification of a sinusoid. Most engineers remember the old full wave rectifier analog power supply.

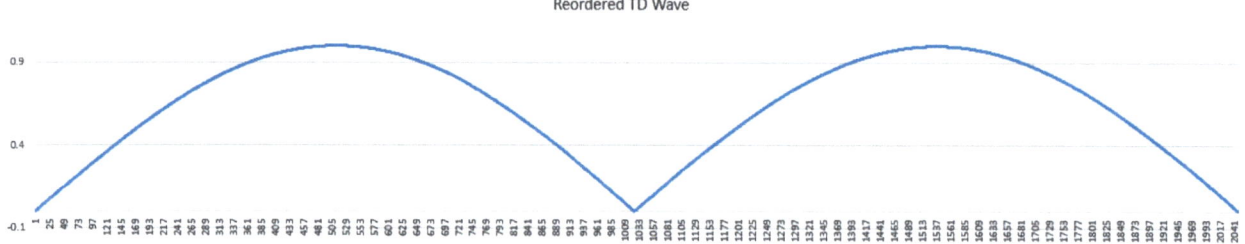

Figure 2 - Time domain display of full wave rectified sine wave

In most cases, a diode bridge converts the negative half cycle of the incoming line frequency to a positive half cycle, utilizing diode steering to get two positive half cycles which yield a positive output voltage. The advantage of this circuit over the half wave rectifier is that the frequency of the line input is doubled, making it easier to filter out the alternations to provide a nice smooth DC output with smaller capacitors and inductors. Linear power supplies are well known for their large size and bulk partly because 60 Hertz is such a low frequency. Airplanes replace 60 Hertz AC with 400 Hertz AC which is easier to filter using lighter inductors. However you would be hard pressed to note the improvement when you hear a PA announcement from the pilot because you can often hear the 800 Hertz ripple on the PA system overlaid on top of the pilot's voice.

Now for the spectrum of the full wave rectified wave above:

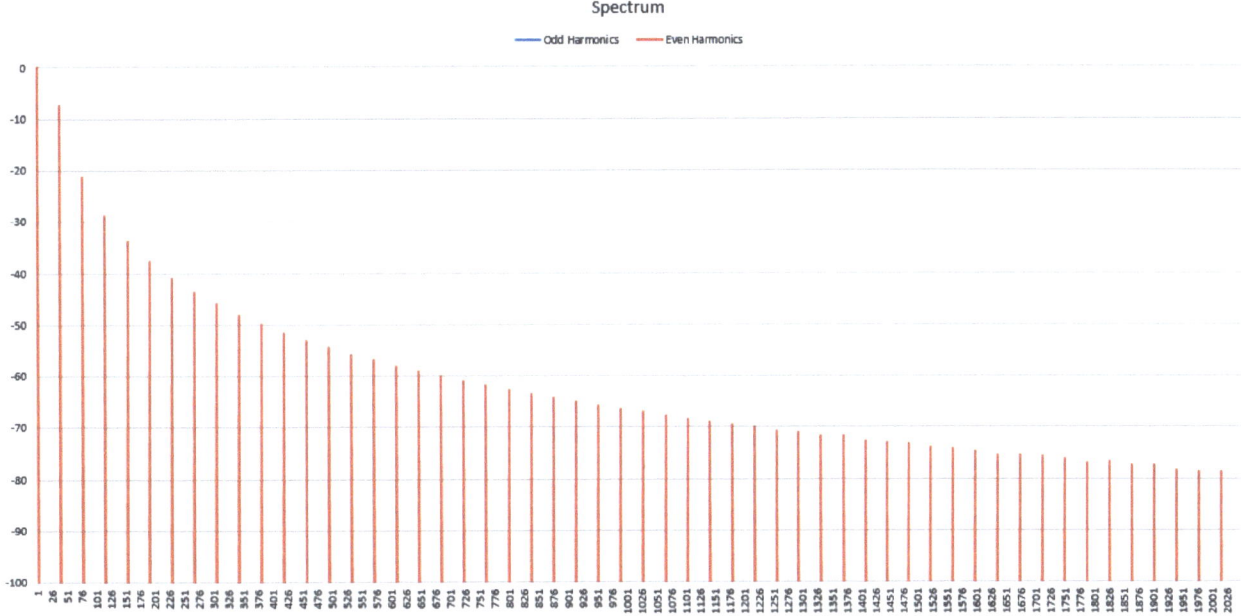

Figure 3 - Spectrum of full wave rectified sine wave

The distortion introduced by rectification doubles the frequency and creates only **even** harmonics (red spectral lines above) of the input line frequency, but that statement ignores the fact that the fundamental frequency is no longer 60 Hertz, but 120 Hertz. The output wave will not have any **odd** harmonics, **odd** relative to 60 Hertz, like 180 Hertz; but, since the real fundamental frequency is 120 Hertz, it makes sense that 180 Hertz would not be a harmonic component of the output signal where the fundamental frequency is 120 Hertz. Note that in order for me to show the above spectrum with **even** harmonics highlighted in red, my color code for harmonic **odd**/**even**ness, I had to lie to the Excel program I used to generate these plots and tell it that the fundamental was 60 Hertz when it clearly was not. Note that the fundamental, 60 Hertz does not even appear in the above spectrum! After DC, the next spectral line to appear is the second harmonic, 120 Hertz in bin 38 (2*19) makes that the new fundamental frequency. So while you can get waveforms that contain only **even** harmonics, it's very, very rare in the real world. This book will concentrate on real world distortion issues, and all of those contain **odd** harmonics along with **even** harmonics in some cases.

Note that I never commit to what frequency bin 19 represents. This time I used it to represent 60 Hertz. Other times I use it to represent 440 Hertz. Other times I intimate that it might be 1 Kilohertz. I don't care about the actual frequency of the fundamental and neither should you. Harmonics are easier to understand by counting frequency bins, where the concept **odd** and **even** harmonics is a lot easier to deal with.

Crossover distortion and **odd** harmonics

People are often confused about **odd** harmonics because they don't always present themselves right away, and most engineers rely on low pass filters to "clean up" their waveforms, either to prevent aliasing or to improve quality of the output signal. Take the case of a Class B amplifier:

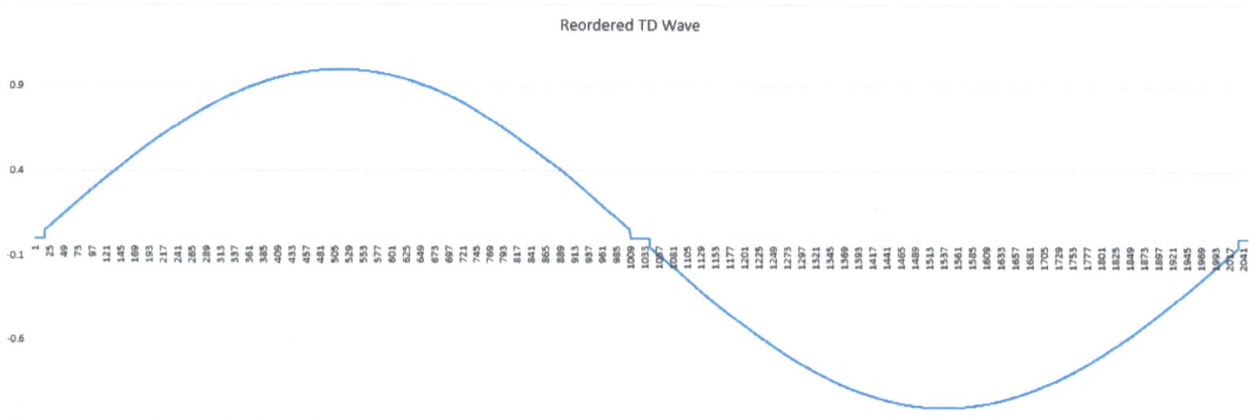

Figure 4 - Time domain display of symmetrical crossover distorted sine wave

Class B amplifiers have a "dead zone" in the center of the transfer function where both the upper and lower transistors (or tubes) are off, so the output goes to zero for some amount of time due to the two diode drops of the un-biased Push-Pull pair.

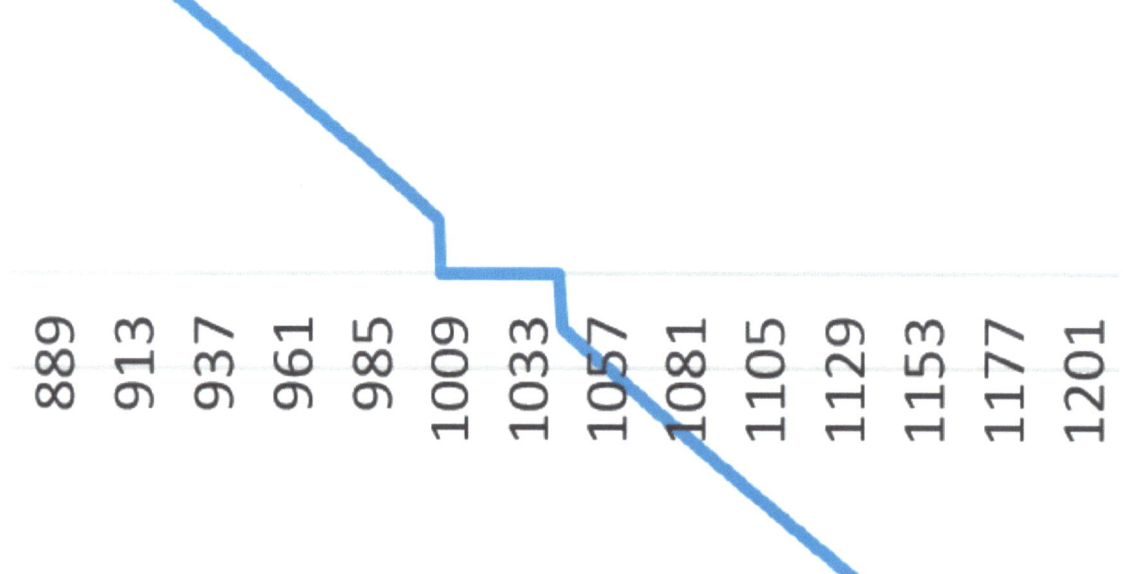

Figure 5 - Closeup of symmetrical crossover distortion

If the two transistors are well matched, the off state will be symmetrical, resulting in only **odd** harmonics below, color coded in blue.

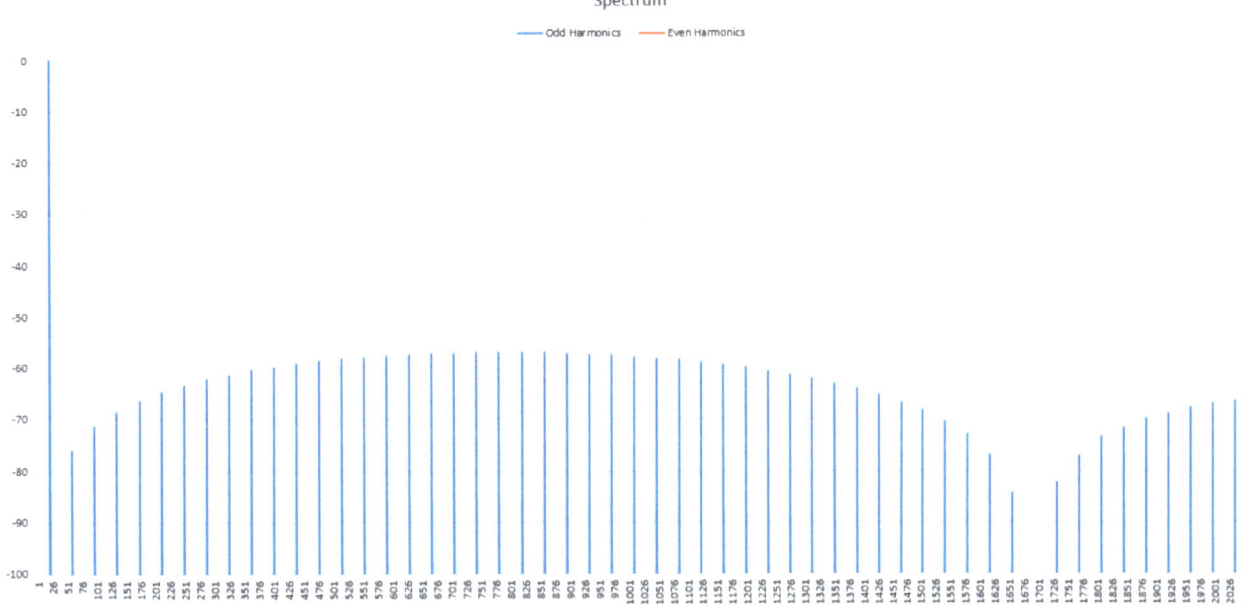

Figure 6 - Spectrum of symmetrical crossover distorted sine wave

However the **odd** harmonics don't start right away. The low order harmonics (3rd, 5th, 7th) are very low in amplitude. It's not until you get to higher order harmonics that they become evident. By then your low pass anti-aliasing filter has probably inhibited their arrival making the **odd** harmonics virtually invisible, especially if you don't look much above the 5th harmonic, which is common in many test specifications. This is what I suspect caused many engineers to believe that symmetrical crossover distortion wasn't such a bad thing, prolonging the lifetime of the Class B amplifier in High Fidelity (Hi-Fi) audio amplifiers. Using low pass filters in the test circuitry hid the deleterious effects of the distortion, along with the assumption that if there were no lower order harmonic products, there wouldn't be any distortion products forthcoming and so there was nothing to worry about.

However, if there is the slightest asymmetry in the transfer function, **even** harmonics will suddenly appear. In the next example I skew the distortion so that most of it is above zero volts.

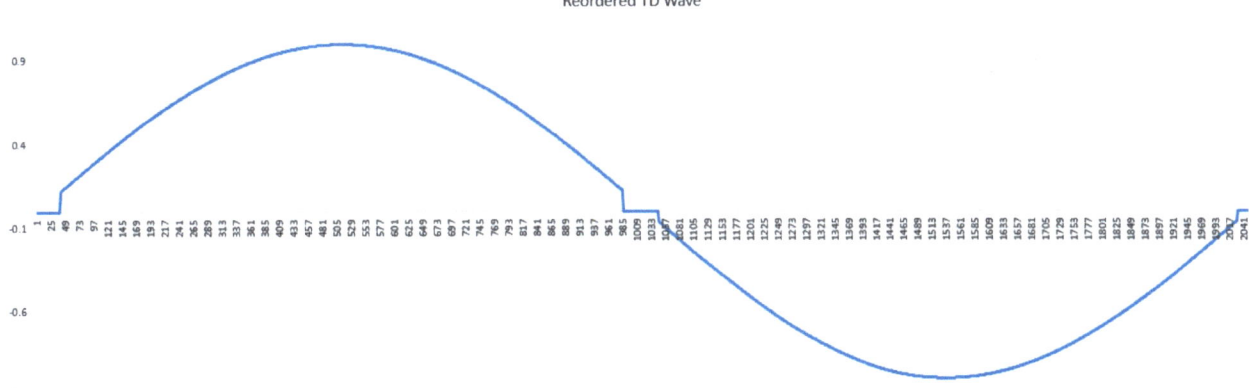

Figure 7 - Time domain display of asymmetrical crossover distorted sine wave

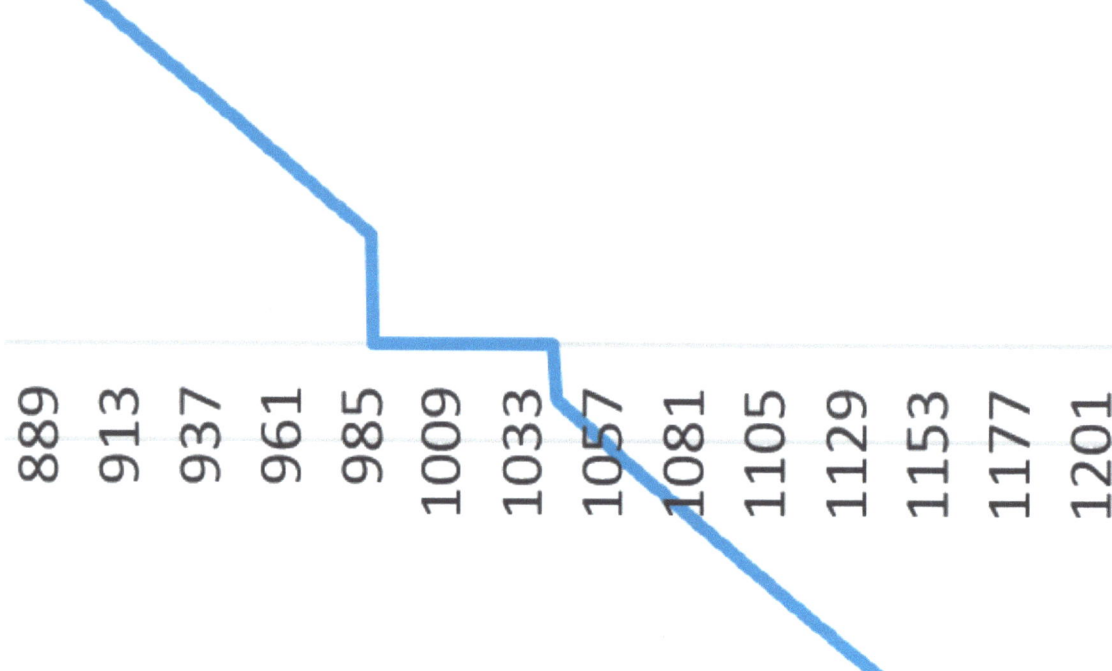

Figure 8 - Closeup of asymmetrical crossover distortion

Note the asymmetrical nature of the distortion, with the negative side exhibiting only 50% of the distortion of the positive side. We perform an FFT and the result looks quite a bit different.

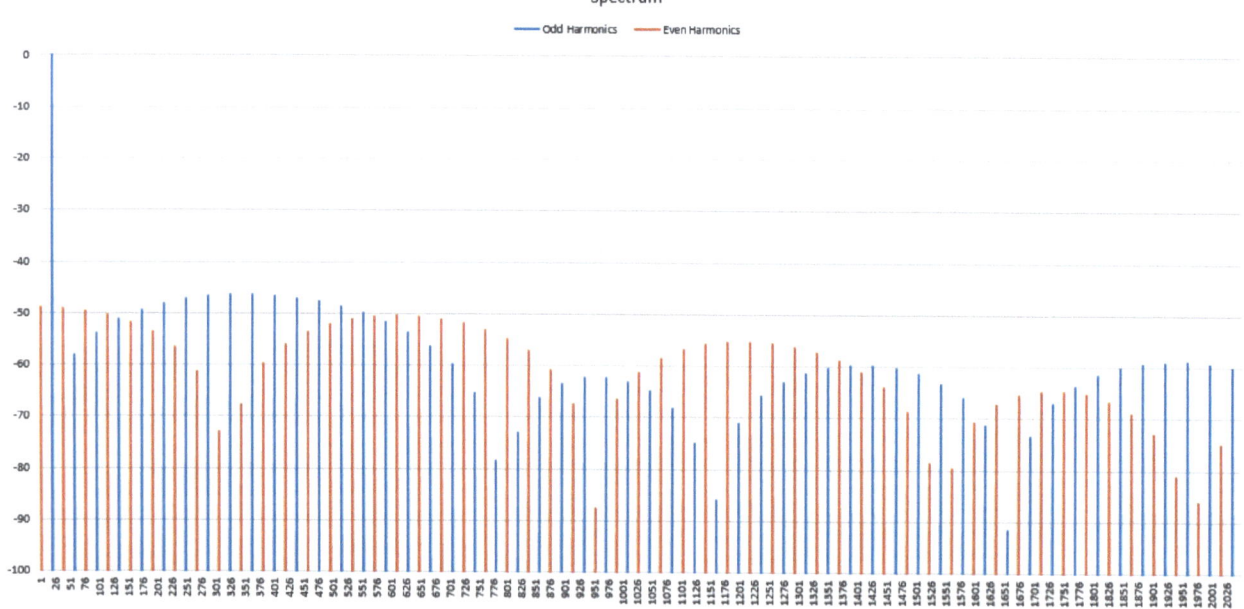

Figure 9 - Spectrum of asymmetrical crossover distorted sine wave

Note that now **even** harmonics appear, and they start with the highest amplitude right at the 2nd harmonic and then decay down towards -100dB. This is unlike what happens with the odd harmonics which start closer to -57dB for the 3rd harmonic and then climb to their peak at much higher order harmonics, usually well after we've stopped looking at harmonics, from the 5th to the 9th depending on the test engineering department guidelines. You might also notice that the blue spectral lines, the odd harmonics, are taller than the even harmonics because the symmetrical distortion area is twice that of the asymmetrical distortion area. Again, a test that stops looking at harmonics after the 5th, 7th or 9th will be fooled into thinking there are no additional harmonics after that based on the **assumption** that harmonics always roll off from the lower order to the higher order harmonics. However, now we know better. Any distortion that happens around the zero crossing causes the odd harmonics to start low, then grow in amplitude the further we get from the fundamental (higher in frequency).

So my statement is confirmed, odd harmonics are always present in any waveform, since the fundamental is always odd. More odd harmonics will appear if there is any distortion and if there is any asymmetry in the distortion, even harmonics will also appear. Even the tiniest amount of asymmetry will cause even harmonics to appear as you saw in the video I made for Applicos.

What is it about asymmetry that makes even harmonics appear? That's actually the wrong question. The right question is this: Why don't even harmonics appear in symmetrically distorted waveforms? The answer here is simple. Since any symmetrical distortion happens in both the positive swing of a sinusoid **and** the negative swing of a sinusoid, being exactly 180 degrees out of phase the even harmonics are canceled out.

The energy is still there, but gets canceled out by the opposite polarity. The only way to see it would be to rectify the signal removing the half cycle that is canceling out the energy of the distortion. But of course rectifying a signal throws a ton of new harmonic content into the mix, as we have seen, making this experiment very unproductive. Later I will show you a wave that confirms this theory, but for now you'll just have to trust me.

So it's not that even harmonics appear only in asymmetrically distorted sinusoids. No matter what the symmetry or asymmetry of the distortion, even harmonics are created; but, if the distortion is symmetrical, the distortion in one half cycle cancels out the distortion in the other half cycle eliminating the even harmonics and leaving only odd harmonics.

Distortion in voltage and time

But even this is not the whole story. The Audio Precision book stated that the symmetry and asymmetry they were referring to was about "zero," which I took to mean zero volts, and that is almost certainly how it was meant. But something else bothered me: The **time** dimension.

Take a square wave with a 50% duty cycle:

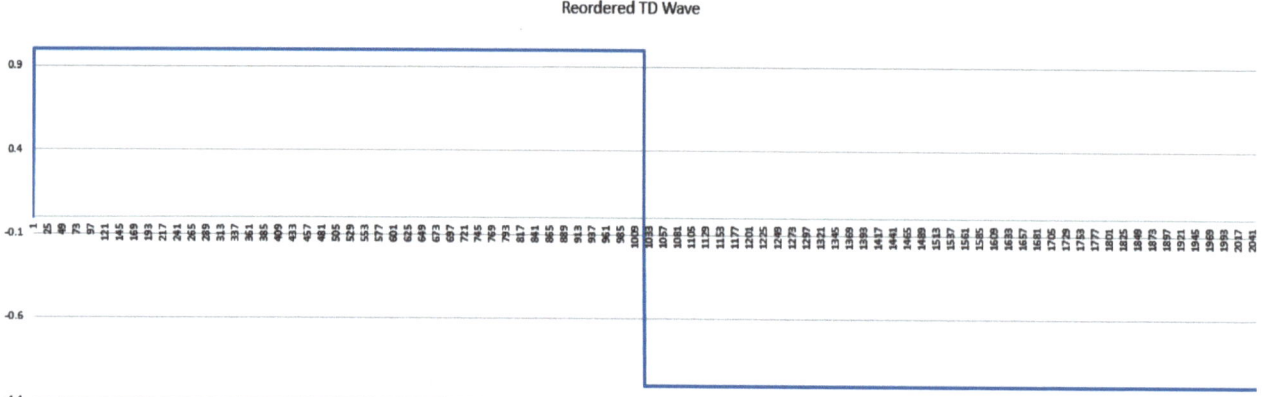

Figure 10 - Time domain display of 50% duty cycle square wave

Do an FFT on it. What do you get?

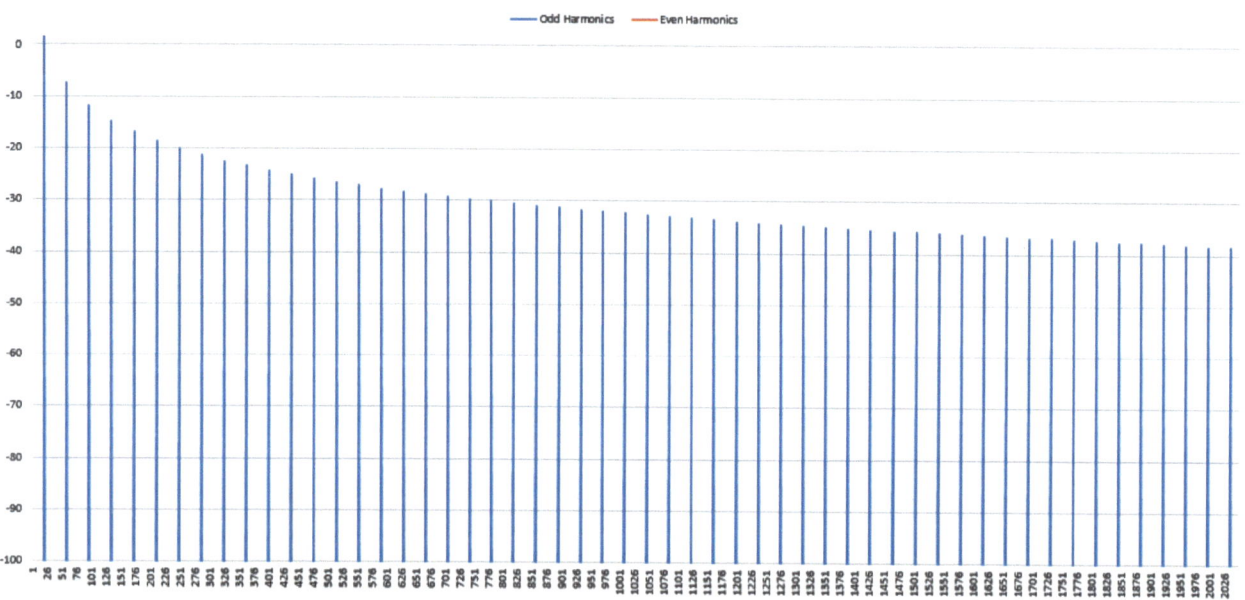

Figure 11 - Spectrum of 50% duty cycle square wave

You get only **odd** harmonics with amplitudes of 1/harmonic# relative to the fundamental. Note that the above graph shows amplitude plotted in dB, so the scale might not match your expectations. So, if the fundamental amplitude is 1V, the third harmonic would be 1/3rd or 0.333333V, the fifth harmonic would be 1/5th or 0.2V, the seventh would be 1/7th or 0.142857V and so on. There will be no **even** harmonics visible, although now we know they are there, but they are canceled out by the symmetry of the distortion (extreme clipping).

Now, change the duty cycle to 51% (or 49%, makes no difference). What happens?

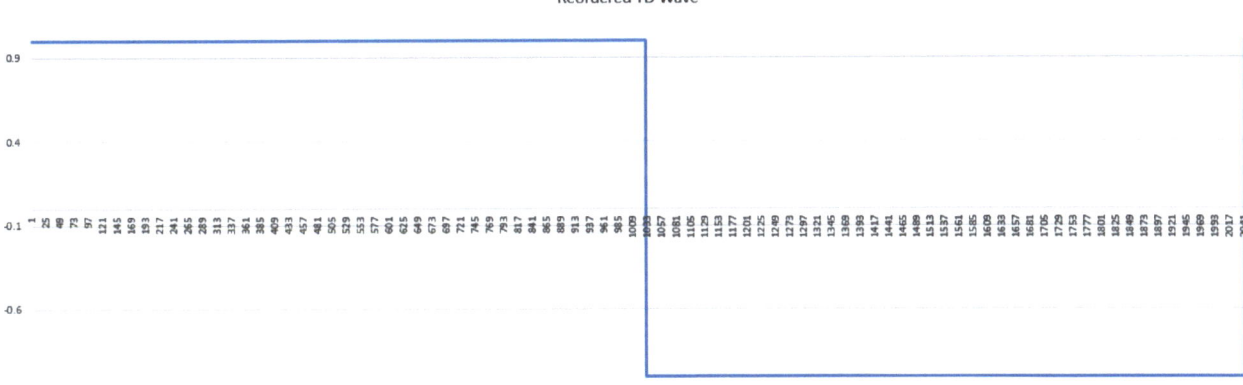

Figure 12 - Time domain display of 51% duty cycle square wave

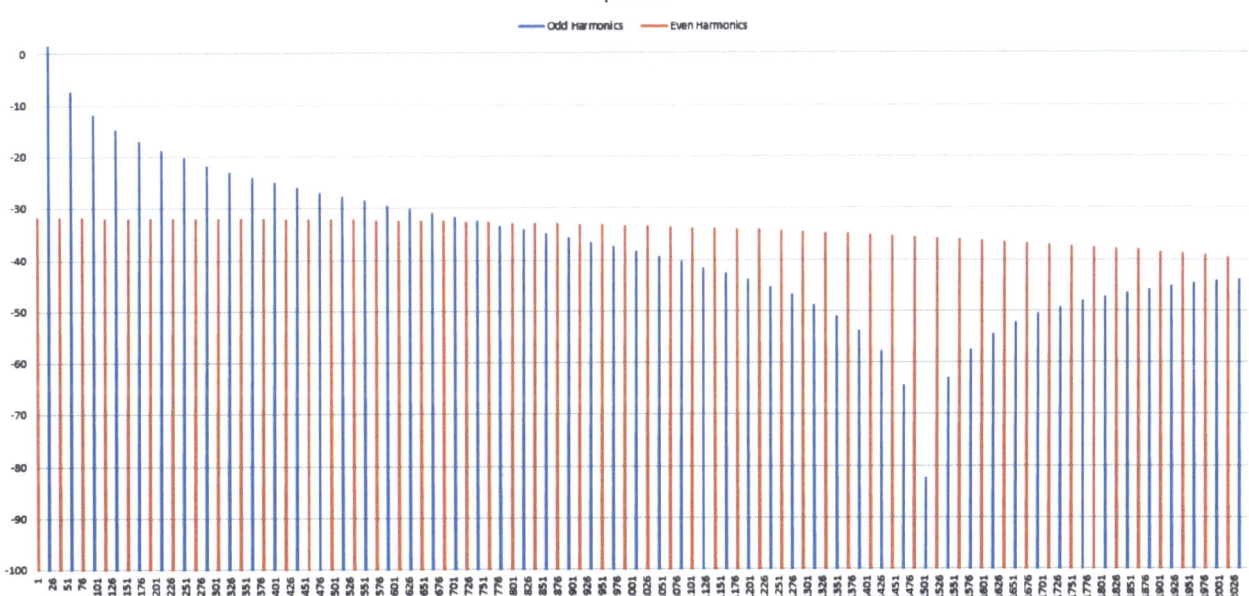

Figure 13 - Spectrum of 51% duty cycle square wave

Suddenly **even** harmonics appear (again, in red). But that makes no sense because the implication in the Audio Precision book (and others) was that **even** harmonics only appear with asymmetries in **voltage**; but, a square wave has only two voltage states, 1 or 0, plus or minus, and whatever the actual levels are, there is no voltage variation in a square wave.

So the asymmetry that allows **even** harmonics to appear does not just apply to voltage, but to *time* as well. In fact, the formula for predicting the harmonic content of a square wave is;

$$\texttt{VHarmonic = abs(sin(π*Harmonic\#*Duty\_Cycle))/Harmonic\#}$$

Duty cycle is calculated as High_Time/Total_Time, so if we calculate the amplitude of the third harmonic relative to a fundamental amplitude of 1.0 we calculate;

$$\texttt{VHarmonic = abs(sin(π*3*0.5)/3) = 0.33333333}$$

But when we calculate the amplitude for an **even** harmonic, like the 2nd;

$$\texttt{VHarmonic = abs(sin(}\pi\texttt{*2*0.5)/2) = 0.0}$$

Try it for a few other harmonics if you like. You will have to agree that this formula perfectly predicts the amplitude of all harmonics for a square wave of any duty cycle. Just make sure your calculator is set to *radians* mode or you'll be misled.

So now we have a formula for calculating the harmonic amplitude based not on voltage perturbations of the waveform, but time. In this formula, the duty cycle is the only variable, and it's in the time dimension, not the voltage dimension. As you will see, this formula can be evolved to calculate the amplitude of any harmonic for just about any type of distortion.

By this time Applicos had grown tired of my *Harmonics Hobby* and refused to pay for any videos I made defending my original video. At least they didn't make me take it down. However, I was now on my own, and I spent months making videos and thinking about why both time and voltage would be responsible for the creation of harmonics.

And then it hit me, *area!* Several years earlier I had been forced to write a *glitch energy* test for a DAC that had a manufacturing defect that produced a large glitch during a major carry transition. To measure glitch energy you multiply time by voltage; and, as any electrical engineer knows, time multiplied by voltage is energy. A very short duration glitch, no matter how large, is trivial compared to a wide glitch of even modest voltage. In the case of RAMDACs, which I had spent a lot of time testing in the past, I had learned a lot about energy. A modestly wide glitch of modestly high amplitude causes what are known as "sparkles" on a Cathode Ray Tube (CRT) because the energy is high enough to excite the phosphors (or not) of the screen making a white or dark spot. Glitches that happen at major carries (0x3F->0x40 for example) were known as "Sparkle Codes," and RAMDACs were extensively tested for Sparkle Codes to make sure that little white or black spots didn't pollute the otherwise beautiful True Color 24 bit RGB images sent to the CRT.

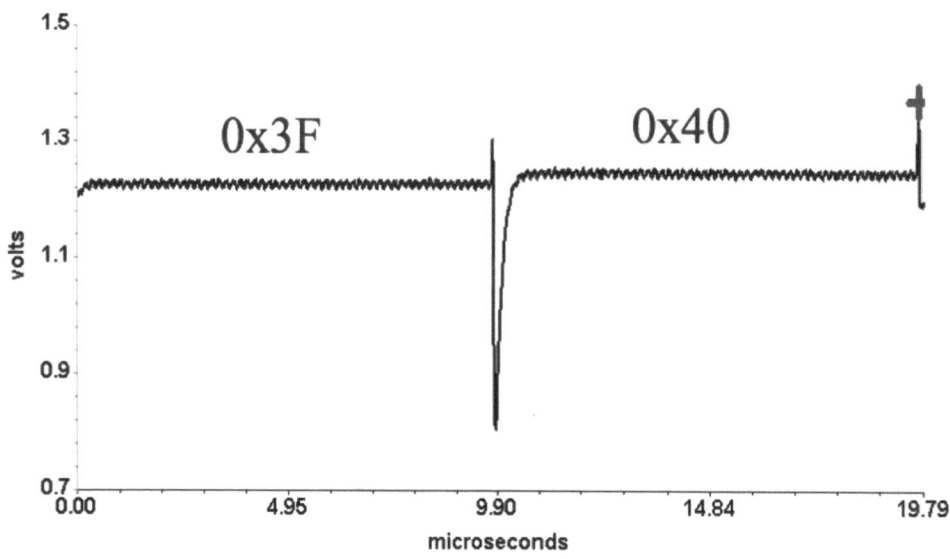

Figure 14 - "Sparkle Code" glitch at 0x3F to 0x40 transition in a DAC

Energy appears in audio too, voltage multiplied by time is energy there too and then you back the audio up with a powerful current source and you get power to drive a loudspeaker which the ear hears and pleases the auditory portions of the brain (or not). When we measure a signal with an FFT what we see is voltage in dB (after processing of course). The harmonics we get from distortion are the frequency domain representation of the *area* of the distortion relative to a pure sinusoidal function.

A pure sinusoid creates a single amplitude in an FFT and in your ears. Now change the amplitude **or** duty cycle of any portion of the sinusoid and you get harmonic distortion proportional to the area of the distortion, minus the even harmonics caused by any distortion that is symmetrical in either voltage or time. It's all about area! The sum of all the harmonic energy in a wave is equal to the area of the sinusoid that is disturbed by the distortion, again, minus any area that is canceled by distortion that is equal in area to the other side of the wave in either the time dimension or the voltage dimension. That is why a 50% duty cycle square wave contains no even harmonics, because despite the fact that the sinusoid is *horribly* distorted, all the distortion is symmetrical in time and voltage, and thus, all even harmonics are canceled out.

Shape matters too

Well, there is another factor. The shape of the distortion is also important and can change the spectral signature of the harmonics created. A good example is the difference between the harmonic content of an INL failure versus a DNL failure.

Integral non-linearity is a very slow perturbation of the transfer function of a device like a DAC.

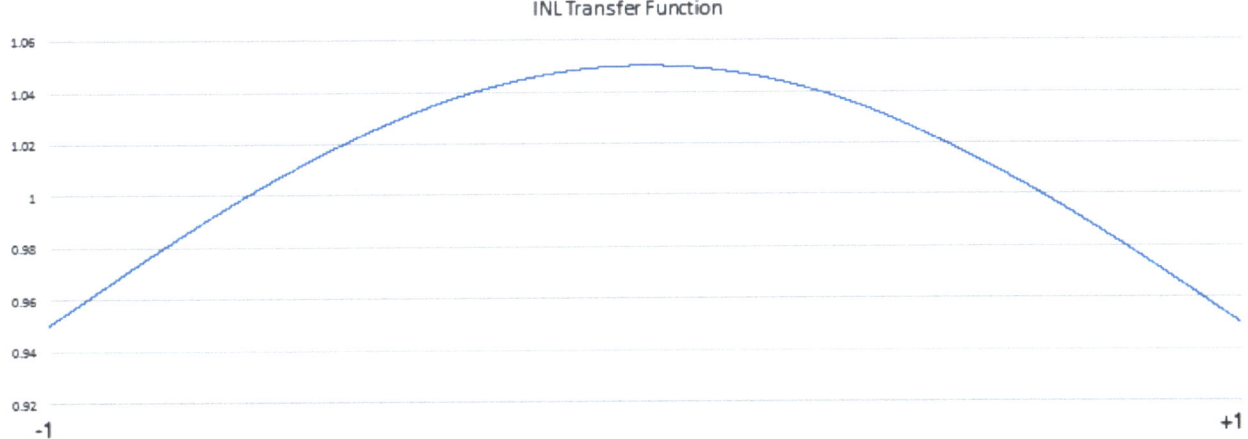

INL Transfer Function

Figure 15 - Simulated INL distortion transfer function

In this simulated high INL transfer function, there is less gain at the positive and negative peaks (at the ends) than there is at the zero crossing (in the center). Because the distortion is so gentle and slow, its impact is almost undetectable just by looking in time domain. In the waveform below, you might notice a little flatness near the peaks, as though the wave has been "squished" from the top and bottom because of the lower gain at edges of the transfer function.

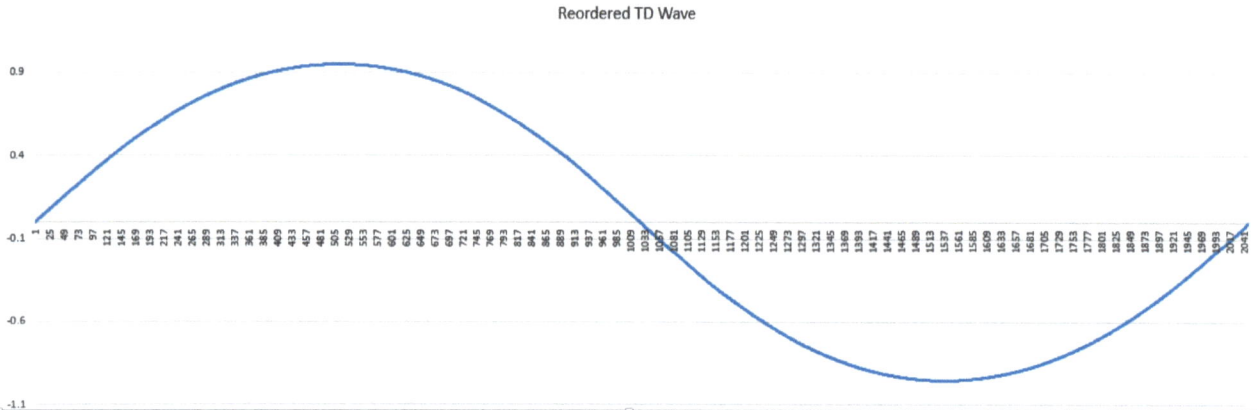

Figure 16 - INL distorted sine wave

Because an INL failure of even one percent (as in this example) impacts every single point in the transfer function, it affects a large area and is therefore quite destructive to the spectral purity of the waveform. However, because it is a very slow curve with no fast transitions, the number of harmonics created is limited to very low frequencies as you can see below:

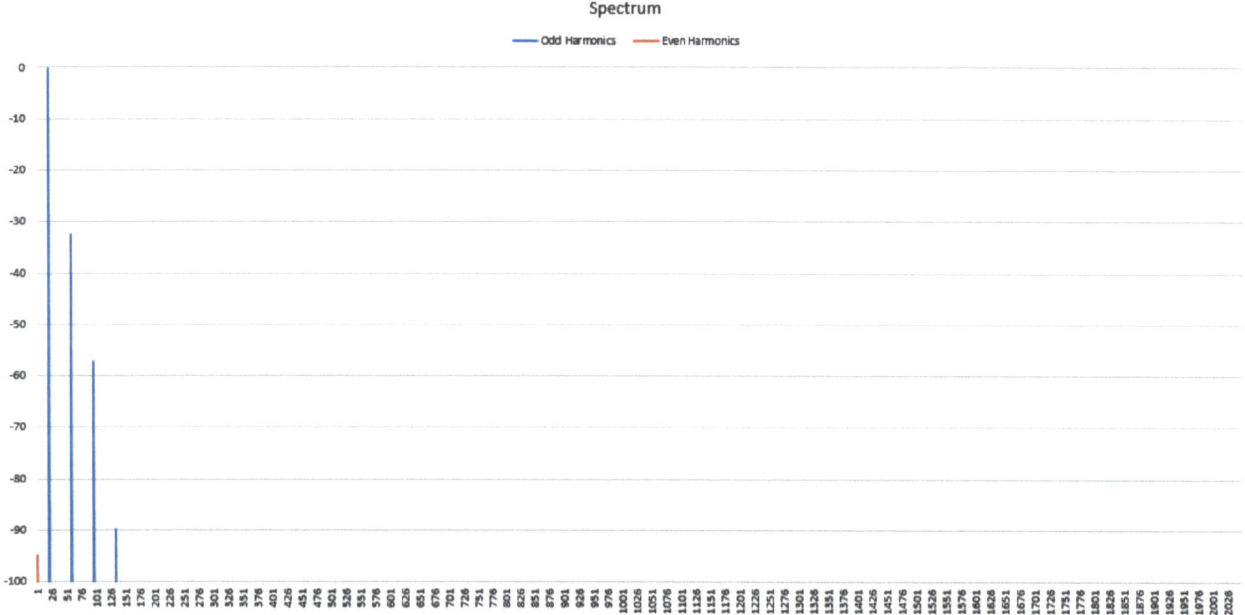

Figure 17 - INL distorted spectrum

Very few harmonics are created by an INL failure of a DAC (or ADC) but their amplitude can be quite large because of the large number of points along the transfer function affected. Note that the spectrum above contains only **odd** harmonics, and all of very low order. The one red bin that pops up is actually bin 0, the DC offset bin. This is because my simulated INL failure transfer function has slightly more area below unity than above, causing a very slight DC offset but is the only **even** numbered bin, there are no **even** harmonics.

The functional opposite of INL, Differential Non-Linearity (DNL) failures are very rapid distortions of the transfer function but their scope is limited to just a few points along the transfer function. Because the total area affected is limited, the amplitude of the harmonics is smallish but the frequencies of the harmonics created can be almost infinite. A DNL error of 4 LSBs on a 16 bit DAC, highlighted in violet below, while small in area (0.05% of the total transfer function area) will have a very fast edge on it, creating many higher order harmonics.

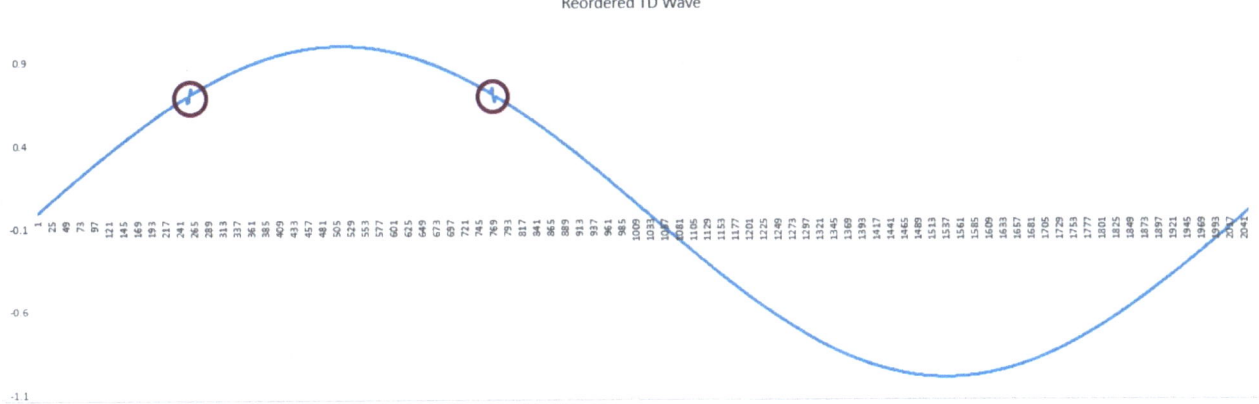

Figure 18 - Asymmetrical DNL distorted sine wave

Because the DNL failure in this example is asymmetrical we suddenly see both odd and even harmonics.

Figure 19 - Asymmetrical DNL distorted spectrum

Notice that the harmonics are of very high order, almost no lower order harmonics appear; and, as we go up in frequency, the amplitude of the harmonics gets higher and higher. Realize that many old time engineers may not understand how destructive DNL failures can be because they often use anti-aliasing filters or other parasitic filters to reduce the amplitude of harmonics above a certain point. Remember that most THD specs never go above the 9th harmonic. What is the THD of the above waveform? For 8 harmonics (2nd through the 9th) the THD is -74dB, but for 50 harmonics it goes up to -66dB. Virtually all engineers assume that harmonics roll off, when clearly they don't in all cases.

Location of distortion in the transfer function

But even this is not the whole story, because it also matters *where* in the transfer function the distortion occurs. If the distortion occurs near the top or bottom of the transfer function, a sinusoid applied against it will be cresting, limiting the total edge height of the transition causing fewer high frequency harmonics.

Here a 1% distortion causes the top of the sinusoid to be lopped off (highlighted in violet).

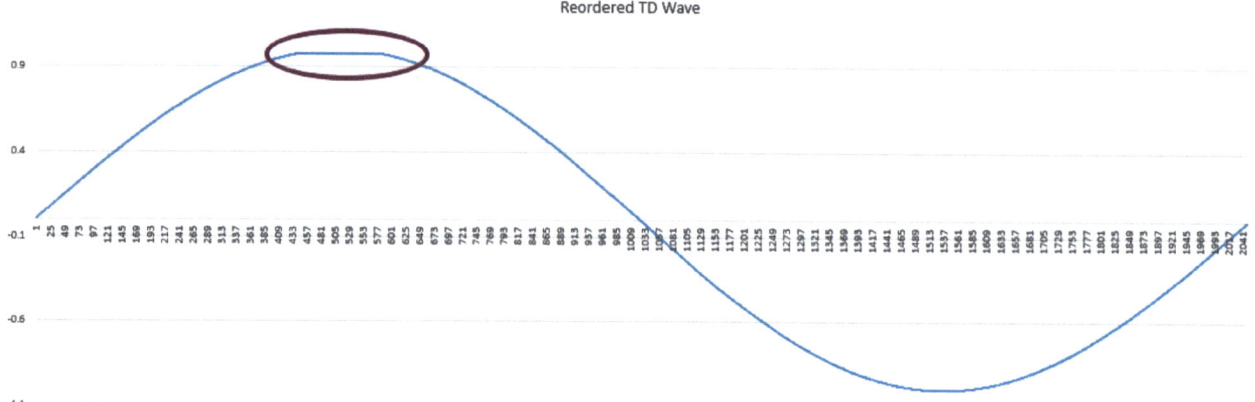

Figure 20 - Asymmetrical limiting distortion at peak on a sine wave

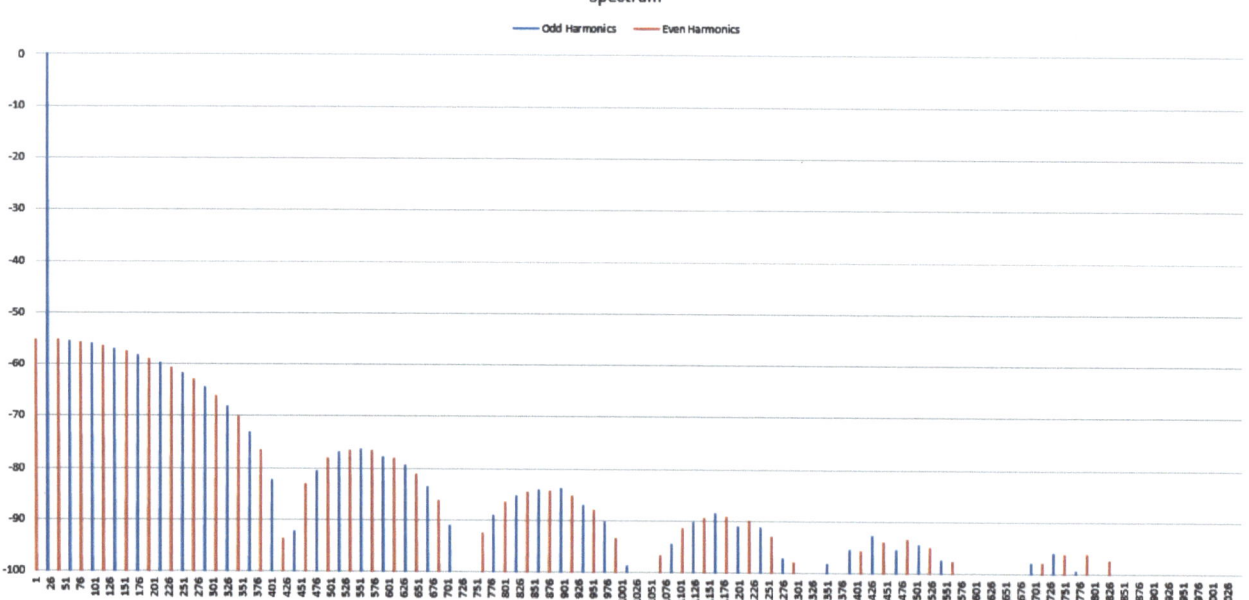

Figure 21 - Asymmetrical limiting distortion at peak spectrum

A fairly large *area* is impacted, so in this example the THD comes out to -47.7dB for 8 harmonics. Extend the harmonic window out to 50 harmonics (includes all audible harmonics for an A4 note used as a test frequency) and the value hardly changes from -47.7dB to -46.6dB. The harmonics drop off quite steeply, addling little as we extend our THD window out because there are no "sharp edges." As I mentioned earlier, most engineers assume that harmonics always roll off like this. But, as I showed earlier with DNL, that is not always the case!

If the distortion happens nearer the zero crossing (again highlighted in violet), where the sinusoid is moving the fastest, the tall vertical edge of the transition creates a lot of high frequency energy, the same as what happens in a square wave with super fast edges. However, the total area of the distortion is minimal, so the THD is quite low, -72dB for 8 harmonics; but, it's much higher if we go out to 50 harmonics to -64dB.

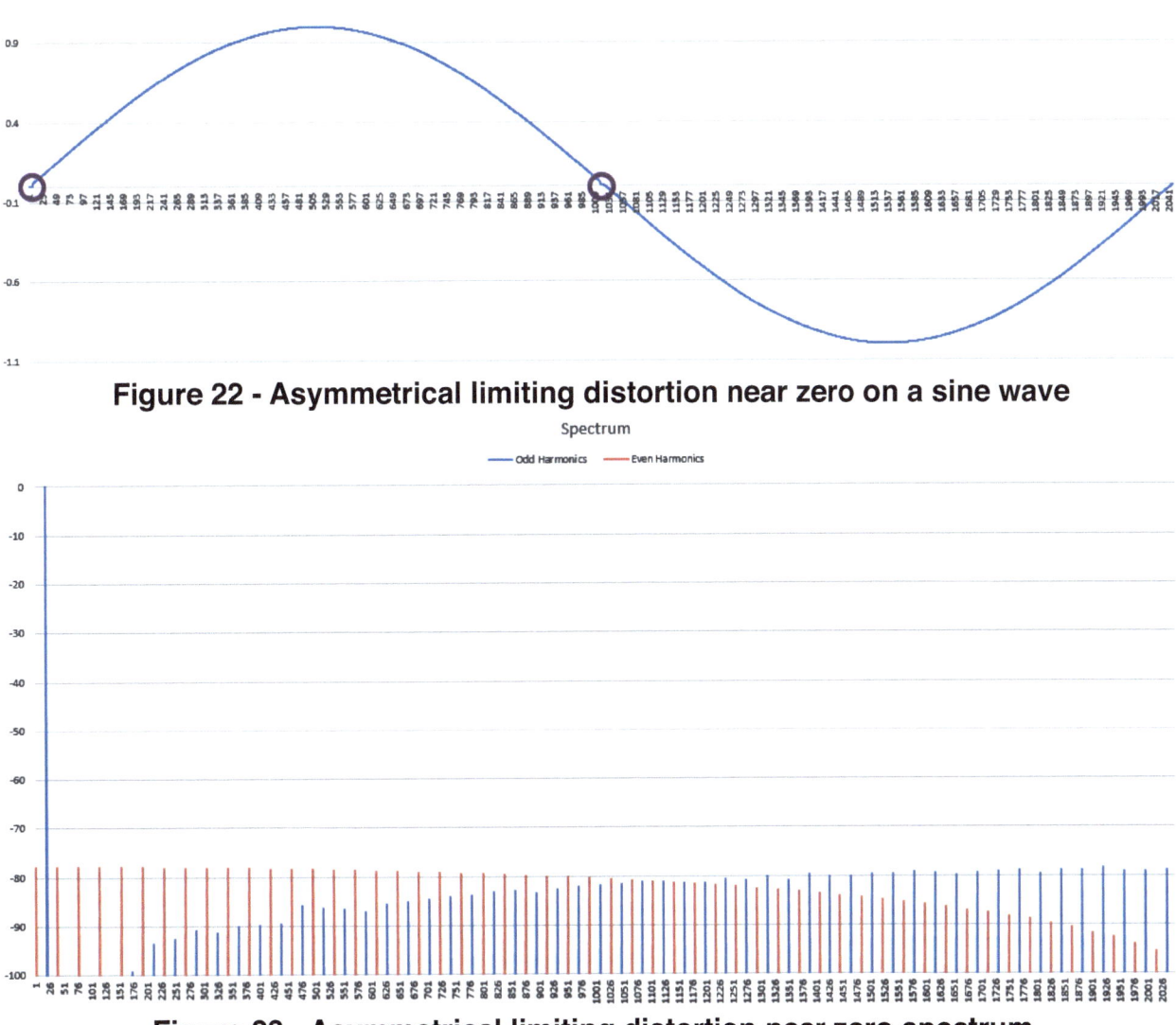

Figure 22 - Asymmetrical limiting distortion near zero on a sine wave

Figure 23 - Asymmetrical limiting distortion near zero spectrum

Notice that a) both **odd** and **even** harmonics appear and b) how the **even** harmonics start out high near DC and yet the **odd** harmonics start off almost non-existent and then rise up as the frequency gets higher and higher. Because the distortion happened near zero on the transfer function, the odd harmonics take a while to build up, and this explains why the THD value goes from a value of -72dB for 8 harmonics to -64dB for 50 harmonics. Who looks out to the 50th harmonic to measure THD? Well, given the fact that distortion near the zero crossing behaves like this, perhaps more test departments should look a little further out.

Now let's try some symmetrical distortion at the zero crossing, again highlighted in violet.

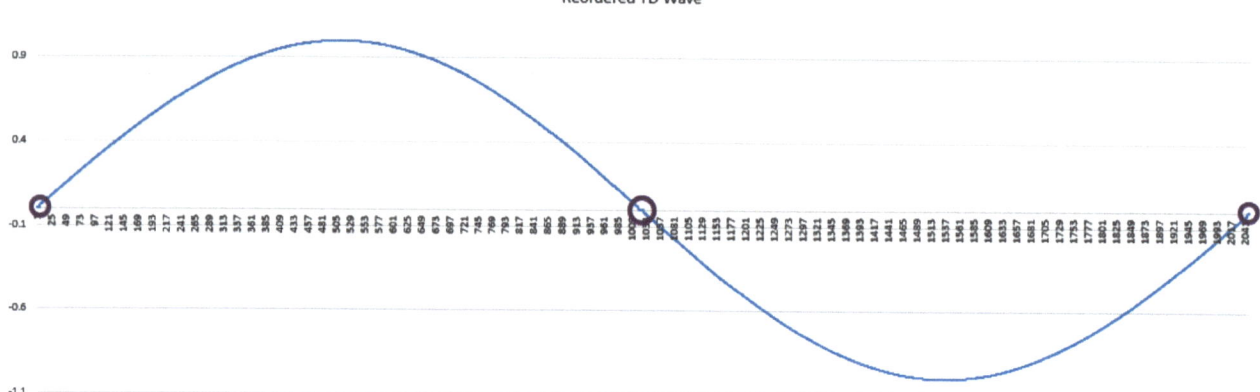

Figure 24 - Symmetrical limiting distortion at zero on a sine wave

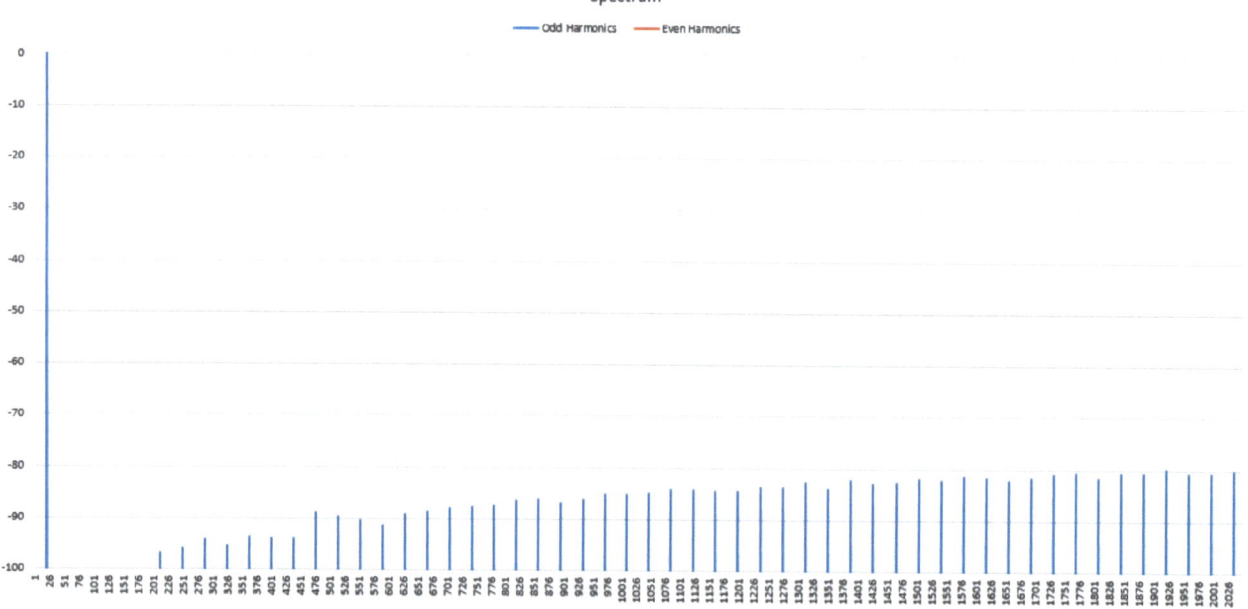

Figure 25 - Symmetrical limiting distortion at zero spectrum

Remember that when distortion is symmetrical there are no even harmonics, making THD very low, -108dB for 8 harmonics; but, because the distortion is at the zero crossing, the odd harmonics ramp up so the THD goes up to -82dB when looking out to 50 harmonics. Who looks at 50 harmonics for THD? Exactly! Perhaps they should!

Note that in all three examples, the amount of distortion is exactly the same; it's just placed in different locations in the transfer function. Not all distortion is created equal. Symmetrical distortion has fewer harmonics because the **even** harmonics get canceled out; the THD will be lower simply because all **even** harmonics are nullified by the symmetrical nature of the distortion. Secondly, distortion near the peaks has a greater influence than distortion near the zero crossing[6].

Distortion impacts the fundamental too

Most of the time we measure the relative harmonic admixture by comparing it to the amplitude of the fundamental, but the value of the fundamental is not static when distortion changes the sinusoid. Some kinds of harmonic distortion decrease the signal strength. Any distortion that removes area from the sinusoid towards zero reduces the amplitude of the fundamental tone. Take this example where a DNL error is simulated: the amplitude of the signal is decreased because a bit fails to flip, so 256 samples have less amplitude than they should; that is, the signal stays closer to zero.

---

[6] https://youtu.be/2GQPqXL8l7Y

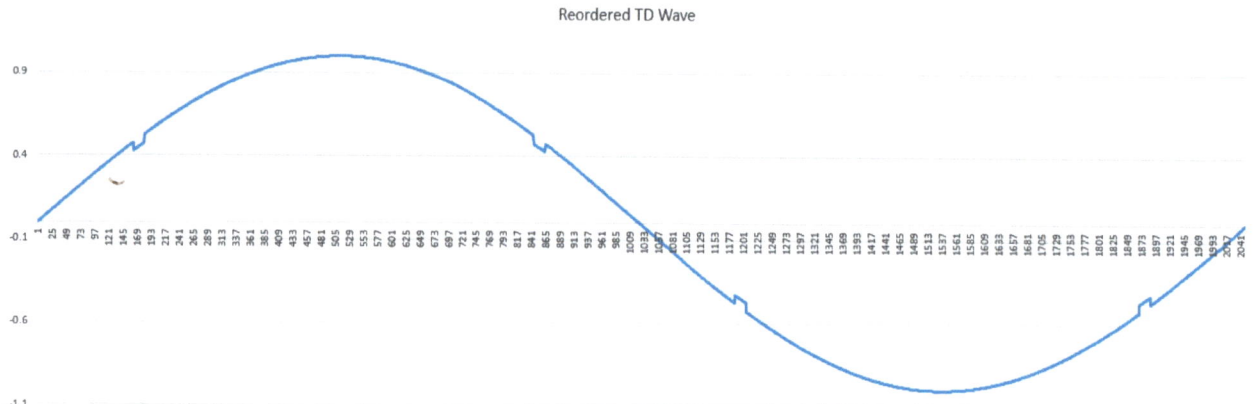

Figure 26 - Symmetrical negative DNL distortion at 0.5V time domain

The spectrum should be pretty simple for you to understand now; we see only **odd** harmonics in **blue** because of the symmetry of the distortion.

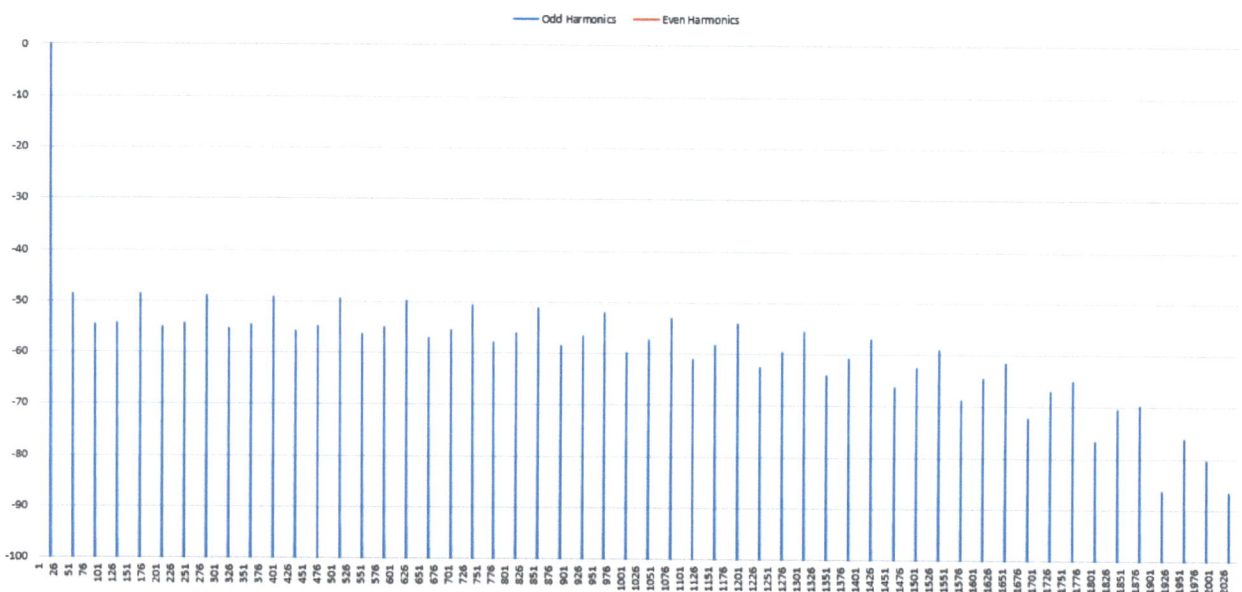

Figure 27 - Symmetrical negative DNL distortion at 0.5V spectrum

In this case, the signal amplitude comes out about 2mV shy of the pure tone expected amplitude of 1.0V peak at 0.998047V. The distortion decreased the signal amplitude because it removed *area* from the sinusoid that the FFT expected to be there. THD measures the relative amplitude of the harmonics vs the fundamental sinusoid, and the value of THD for this example comes out to -44.7dB for the first 8 harmonics and -38.5dB for 50 harmonics.

Now let's try reversing the polarity of the distortion. This time we will *add* signal by flipping this bit in the other direction. Here is the time domain version:

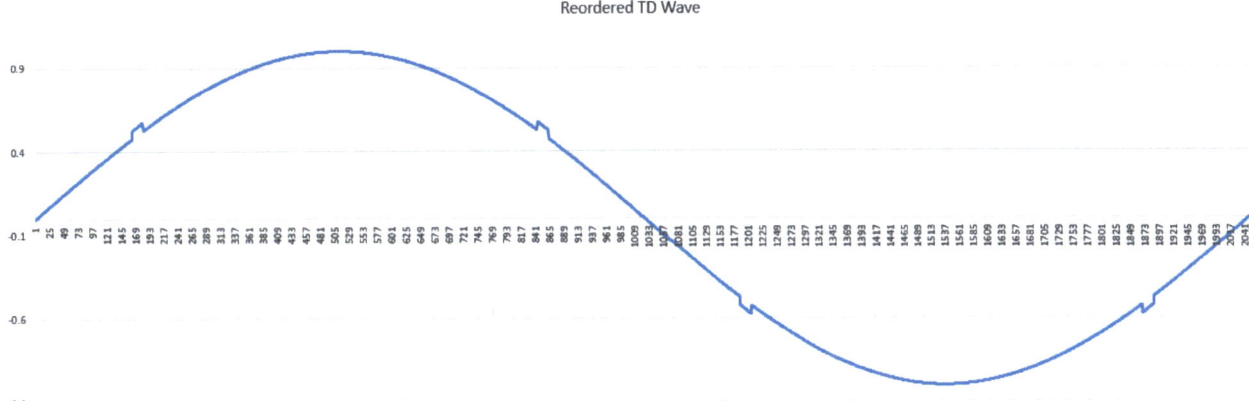

Figure 28 - Symmetrical positive DNL distortion at 0.5V time domain

In time domain you can see how it looks like the signal is trying to "grow"—be a little bigger in the middle of the slope than it should be. This adds signal to the sinusoid, and the FFT takes notice.

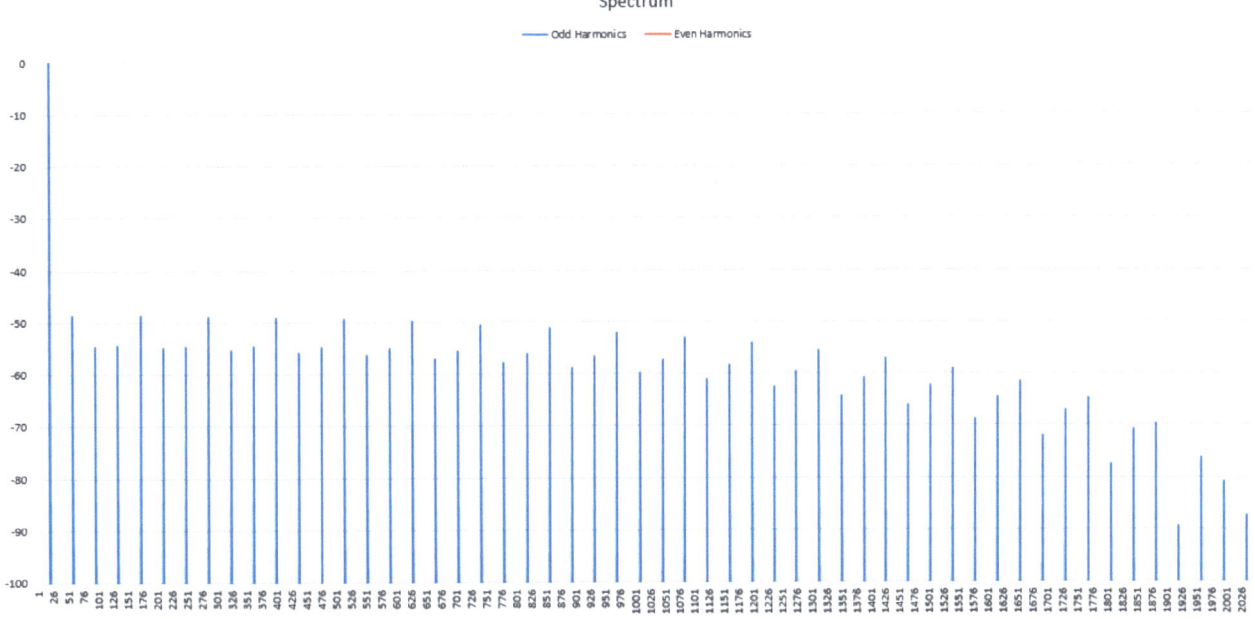

Figure 29 - Symmetrical positive DNL distortion at 0.5V spectrum

It's hard to see since everything is scaled in dB, but you should be able to tell that the two spectra look almost exactly the same. The numbers don't lie though; in this case, the signal amplitude is 2mV higher than it should be at 1.001953V. Despite the fact that the amplitude of the fundamental went up by 2 millivolts (due to the extra area), the THD numbers show that the magnitude of the harmonics also increased to make up for the difference showing a THD of -44.7dB for 8 harmonics and -38.5dB for 50 harmonics— exactly the same as before with the negative DNL example.

It is rather surprising that the amplitude of the fundamental should change, but it does, and so does the harmonic content. It makes sense if you realize that the signal amplitude is the delta from zero. Distortion that removes *area* from our signal, as in our first example, decreases the amplitude of the fundamental sinusoid, but it also *slightly* decreases the harmonic amplitude since the distortion moved the wave towards zero. In the positive DNL example, the distortion increases the area of the signal, but it also increases the area of the harmonics too.

The more area that the signal occupies, the more energy it contains. That is one downside to overdriving an audio amplifier. If you drive a large enough sinusoid into an audio amplifier it causes the amplifier to saturate at both the top and the bottom[7]. You have now created a square wave. A square wave has a lot more area than a sine wave, and the extra energy (area) is translated into extra power, which could exceed the spec of the amplifier and heat sink, causing the amplifier to overheat and burn up.

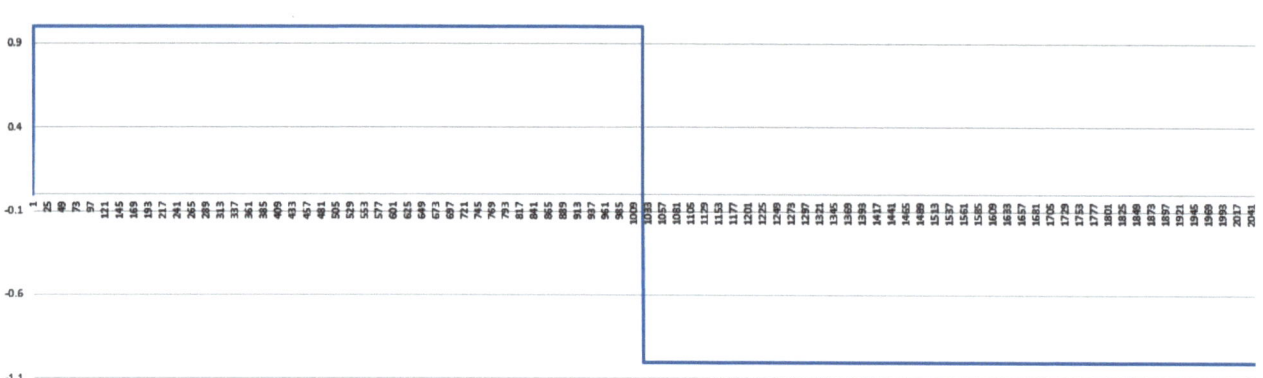

Figure 30 - Square wave created by overdriving a linear transfer function

This 50% duty cycle square wave has 272.5mV more amplitude in the fundamental bin than the sine wave it is derived from.

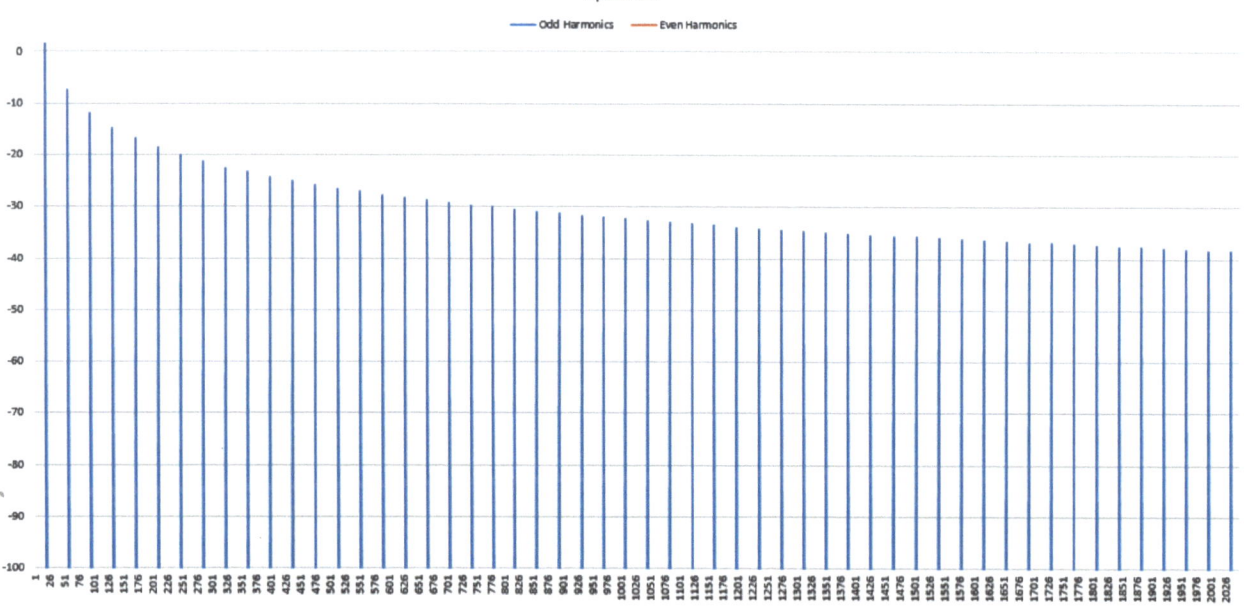

Figure 31 - 50% duty cycle square wave spectrum

7 https://youtu.be/yxc-GCmYKXE

That's an additional 2.1dB for the theoretical 1 volt peak signal. Now add all the harmonics that are created. A 50% duty cycle square wave is loaded with harmonic energy because of the extra *area* of the square wave over a sine wave. The THD of this wave is -6.5dB, meaning that a third of the energy in the wave is expended in harmonics. So now the signal is 2dB over the maximum expected power for a low THD waveform (a pure sine wave); but, in addition, the odd harmonics show up and further overload the amplifier with 50% more power than we expected.

Now, get away from 50% duty cycle, and suddenly the even harmonics show up. Does that change the situation? Surprisingly, no. Remember that all distortion creates both odd and even harmonics, but symmetrical distortion allows the positive side to counter the negative side and squash the even harmonics down. At 51% duty cycle suddenly the even harmonics appear, but at the expense of the odd harmonics.

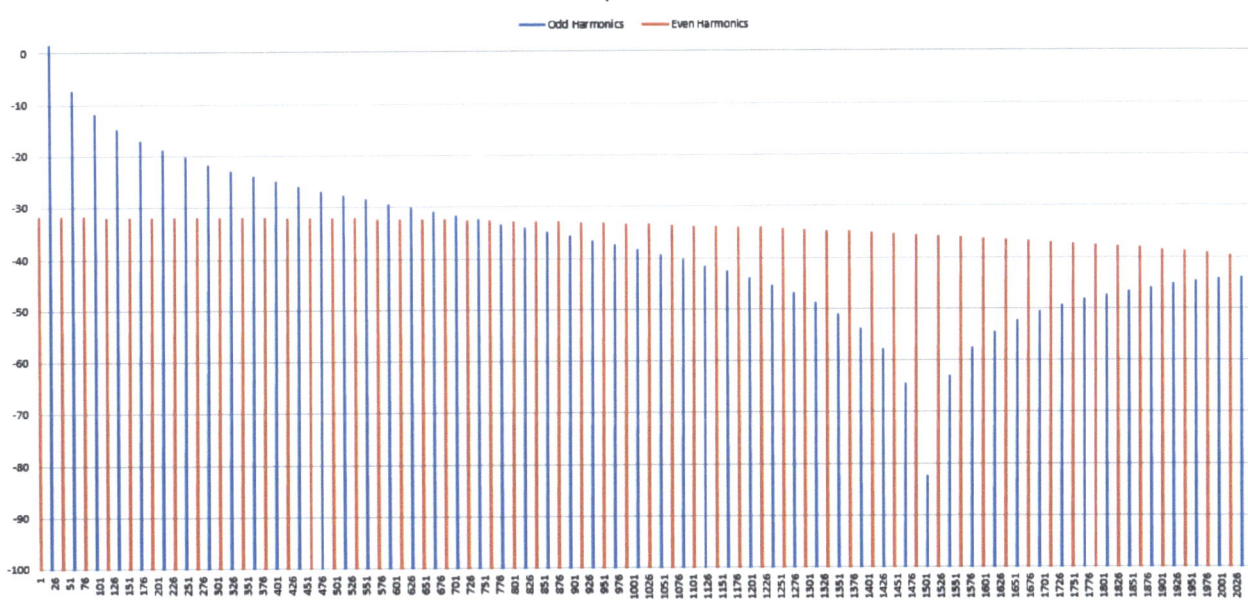

Figure 32 - 51% duty cycle square wave spectrum

Notice that although the even harmonics suddenly appear, the odd harmonics are reduced in amplitude leaving the THD unchanged at about -6.5dB. Let's face it, a square wave has the most energy (area) you can put into an amplifier, and not every amplifier can take it. They often shut down when overwhelmed with maximum energy.

I've actually seen this happen to an audio power amp with thermal shutdown protection during a party. The host was trying to make sure everyone attending could enjoy the music, and the power amp kept shutting down. He touted it as a feature, but nobody else thought that music playing, then shutting off, then coming back on after a few seconds of cooling, was very conducive to the festive mood.

Driving too much voltage into an amplifier causing it to clip increases the total area (energy), and the power amp provides almost unlimited current to drive the speakers, causing a horrific THD and overheating the amplifier to the point of thermal failure.

Now, don't be so naive to think that the only waveforms that will pass through your amplifier will be sinusoids. The music industry tends to lean towards Phil Spector's concept called the Wall of Sound[8]. Multiple instruments are orchestrated to completely fill the speakers with maximum audio content, but without clipping, so the instruments are still identifiable (brass, strings, vocals, percussion, etc) but they are so phased that the listener is overwhelmed by audio content. Phil Spector had it right, the Wall of Sound has a visceral psycho-acoustic appeal that brings in music lovers, convincing them to buy albums. But at a cost: the amplifiers must be able to handle this energy rich spectrum.

The point is, don't expect only sine waves. If you design an audio amplifier to handle a square wave, you'll never have to worry about the thermal protection kicking in and shutting down the stereo or other apparatus at an inopportune moment. Area is energy; plan for maximum area, and you can't go wrong.

---

[8] https://en.wikipedia.org/wiki/Wall_of_Sound

Chapter 2
Waveforms - a zoo or a continuum?

Read any book on electronics and you'll find a section describing various waveform types. There are sine waves, of course. There are square waves, sawtooth waves, triangle waves, rectangle waves, ramps, pulses, rectified waves, a whole Noah's Ark of waveforms that are unrelated. According to these experts, each waveform needs to be studied as its own zoological form, like a horse is different from a dog, or a bird is different from a turtle.

This claim is so far from the truth it's almost laughable. It's almost like taking a Rubik's Cube, twisting one side and now that the color pattern is different claiming it's not a Rubik's Cube, it's another thing entirely!

Fourier told us there is only one kind of waveform, it's a sine wave, and any other waveform is a distortion of a sine wave by adding other sine waves, most often harmonics in some regular mix with some particular phase relationship.

I've already shown how you can create a square wave[9] by overdriving a sine wave to the point where there is no voltage information other than the top and the bottom, 0 or 1, +5V or 0V, whatever—a square wave has only two states. The harmonic content is all **odd** harmonics with amplitudes of 1/harmonic#, so that with a 1 volt peak fundamental, the third harmonic is 0.333333 (1/3) volts peak, the fifth harmonic will be 1/5th of a volt or 0.2 volts peak, the seventh harmonic will be 1/7th or 0.142857 volts peak and so on.

Then I showed you how a "rectangle" wave that is not symmetrical in time differs from a square wave only in the fact that **even** harmonics become part of the spectrum. The more asymmetrical (in time, hence area), the more the **even** harmonics come up and then a beautiful dance begins with the **odd** and **even** harmonics[10] forming wonderful mathematical patterns as they zig and zag past each other. The mathematical purity of this dance tells you that something is afoot, and it would be best not to ignore it.

In fact, the formula we saw in chapter 1 told you all you need to know about these waveforms. You may remember it was;

$$\texttt{VHarmonic = abs(sin(}\pi\texttt{*Harmonic\#*Duty\_Cycle))/Harmonic\#}$$

So in one fell swoop we have grouped all "square-ish" waveforms, square waves, rectangular waves and pulses into a single species with a single formula. The only way that a pulse wave differs from a square wave is in duty cycle. An infinitely narrow pulse, called an impulse, has a really interesting spectrum.

---

[9] https://youtu.be/yxc-GCmYKXE

[10] http://youtu.be/cKM5Z9b1bos

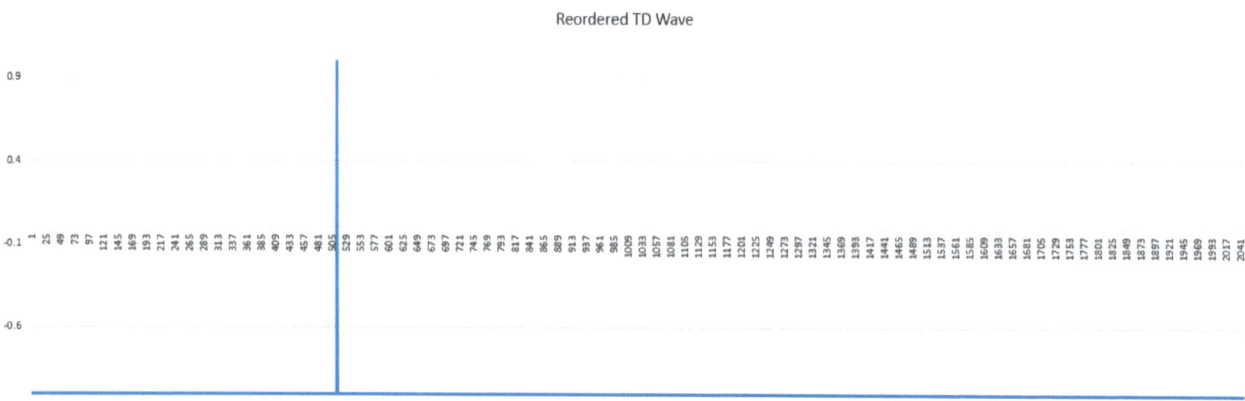

Figure 33 - Single point impulse in time domain

The perfect impulse consists of a single point high amid a field of all zeros. If more than one point is high, the spectrum will be compromised. If done right however, you get a spectrum that looks like this:

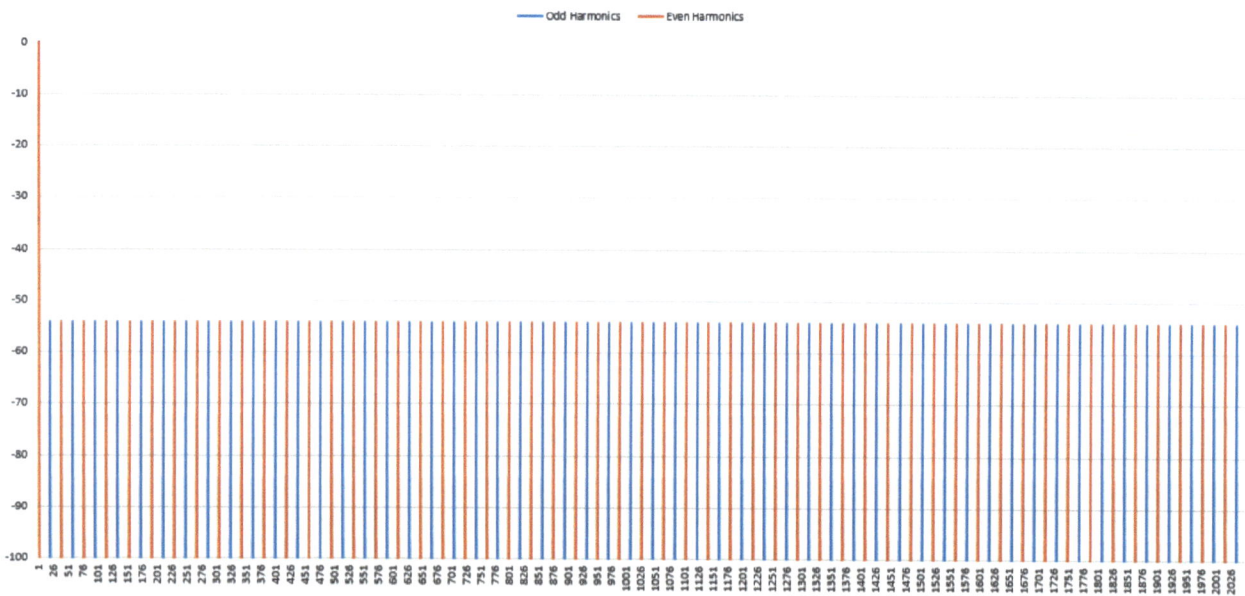

Figure 34 - Single point impulse spectrum

This spectrum is so pretty it is almost hard to believe. It's a perfectly flat array of **odd** and **even** harmonics. It's so perfect that many engineers have tried to use an impulse to test filters. Drive a filter with an impulse, capture the output, do an FFT, and you now have the spectral response of the filter in one shot. The only problem is that because the impulse has such a small area, it lacks energy, and thus the signal-to-noise ratio is severely compromised. There is a better way, but it is outside the realm of our current topic, so I will simply give you the link to my article[11] to read at your leisure.

---

[11] https://www.evaluationengineering.com/its-all-in-the-noise

Why do we get a flat spectrum from an impulse? Do the math on the formula I gave you earlier using a duty cycle of 0.0000001. You get the same tiny number for every harmonic you calculate. The flatness is a good thing, but the tiny amount of voltage in each spectral bin makes using this technique in anything but a purely noise free environment problematic at best.

You might think to yourself, "Well, the best way to fix that is to make the impulse wider, which gives it more energy improving the signal to noise ratio!" Good luck with that, let's just see what happens.

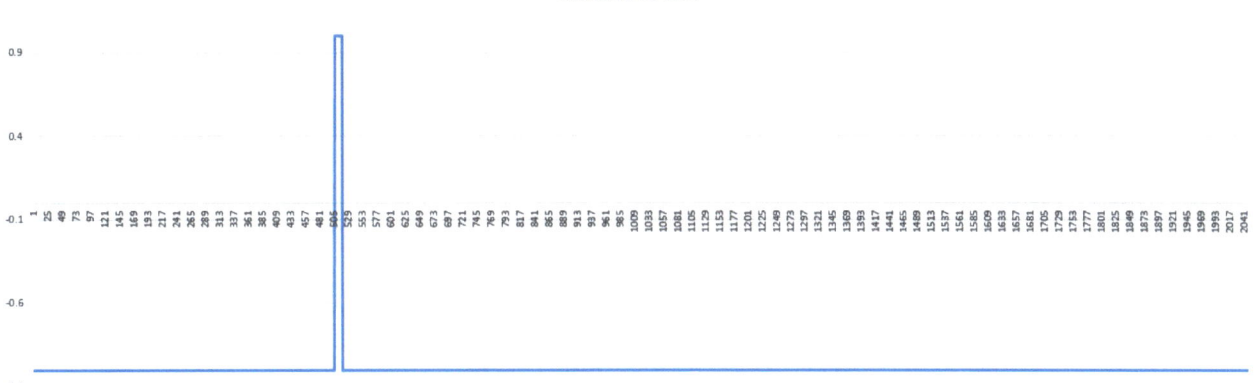

Figure 35 - Thirteen point impulse in time domain

While the time domain version looks OK and clearly has more energy, it's best to see what the spectrum looks like.

Figure 36 - Thirteen point impulse spectrum

This spectrum is starting to roll off, which makes it unsuitable for use in testing filters. Why is the spectrum rolling off? Check your formula with a duty cycle of 13/2048:

$$\texttt{V}_\texttt{Harmonic} = \texttt{abs(sin(}\pi\texttt{*Harmonic\#*Duty\_Cycle))/Harmonic\#}$$

The new-found energy is directed into the lower frequency harmonics, leaving the higher harmonics at the lower levels. At least that's one way of looking at it.

We have already seen what happens when the duty cycle deviates from 50%. The nice curvaceous patterns that we see in the frequency domain are pretty, but they aren't good for much, certainly not filter testing. However, it has been done. Several years ago I collaborated on an article for Evaluation Engineering[12] on how to use a plain old square wave to test filters. It works pretty well, but we had problems with some specific applications.

Now that you know something about the difference between square waves, rectangular waves and pulses, think about the spectrum of a pulse width modulator (PWM). Imagine all the wild spectral patterns generated by a PWM device such as a Class D amplifier. When you drive a Class D amplifier out to a speaker, what happens? The very low bandwidth of a highly inductive speaker coil along with a low pass filter removes most of the harmonics and turns that pulse width modulated signal into a sinusoid. That's the beauty of Class D: the speakers, as highly inductive audio transducers provides some of the low pass filtering to remove the plethora of harmonics to make any duty cycle square or rectangular wave into a sinusoid. But now you are counting on a parasitic filter (the speaker) to clean up a very ugly spectrum. Be careful, the quality might not be as good as your instruments lead you to believe[13].

To sum up, there is no such thing as a pulse wave, a rectangular wave or a square wave. As Fourier proclaimed many years ago, there are only sine waves, nothing else. Every wave is just the accumulation of sine waves at various frequencies and phase relationships, most often harmonically related. Does this apply to triangle waves, sawtooth waves and ramps too? Take a wild guess, and read on.

Triangle Waves

Cathode Ray Tube (CRT) oscilloscopes and televisions were made possible by the use of a sawtooth wave to drive the horizontal sweep circuits that caused the electron beam to move from left to right across the screen. The sawtooth wave was created by a current source driving a capacitor which creates a linearly rising voltage. At some threshold, a comparator trips and turns on a switch that drains the voltage off the capacitor, the switch turns off, and the linear charge curve starts again. Without the sawtooth wave, the oscilloscope and television could not exist.

How is a sawtooth wave related to a sine wave? The answer is again, very simple. The formula below shows how various sinusoidal harmonics can be added to create a sawtooth or triangle wave:

$$\texttt{V}_\texttt{Harmonic} = \texttt{(abs(sin(}\pi\texttt{*Harmonic\#*T}_\texttt{Rise}\texttt{/T}_\texttt{Total}\texttt{))/Harmonic\#\^{}2)/}$$

$$\texttt{sin(}\pi\texttt{*T}_\texttt{Rise}\texttt{/T}_\texttt{Total}\texttt{)}$$

---

[12] http://www.danbullard.com/dan/low_pass_testing_cover.html

[13] http://www.eetimes.com/document.asp?doc_id=1274731

This formula gives the amplitude of each harmonic relative to the fundamental (which scales to a value of 1.0 for the fundamental tone) for a triangle wave, sawtooth wave, or ramp based on the ratio of the rise time versus the total time. We can't use the term "duty cycle" for triangle and sawtooth waves but we can talk about TRise to TTotal ratios (TR/TT). For example, here is the time domain plot of a perfectly symmetrical triangle wave (50% TRise/TTotal ratio):

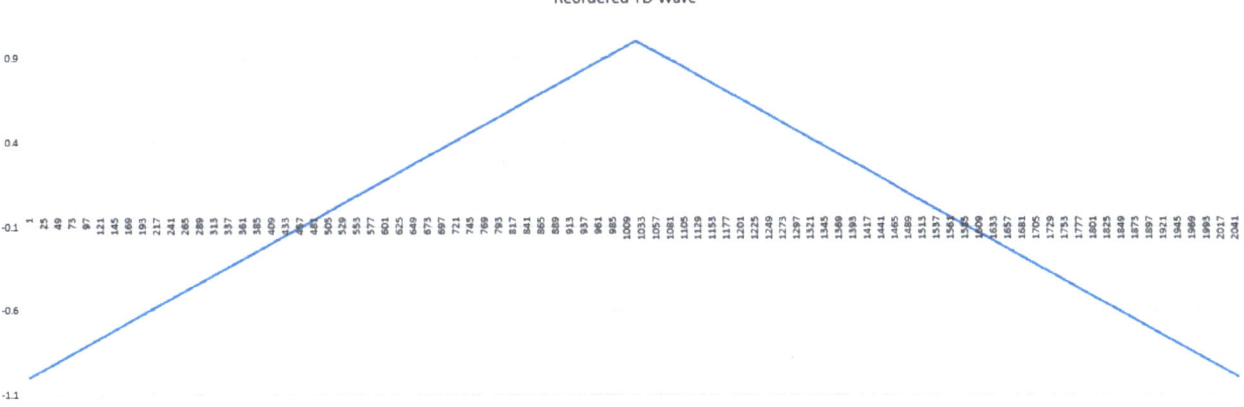

Figure 37 - 50% TR/TT triangle waveform

This waveform is perfectly symmetrical in time *and* voltage, so you can expect that the spectrum will have plenty of odd harmonics and no even harmonics.

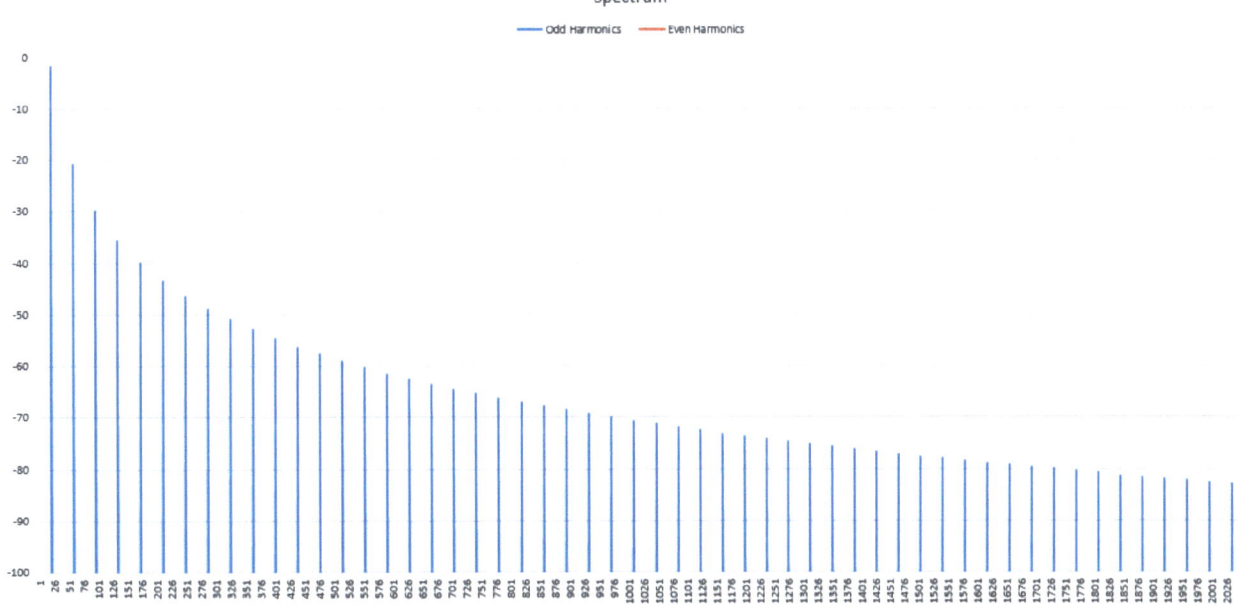

Figure 38 - 50% TR/TT triangle wave spectrum

Note that there are no even harmonics; even the DC bin is empty. That's because the signal starts at -1V, goes to +1V, then back down to -1V; and, the total area above zero is identical to the total area below zero. Also note that, although it might be hard to see in a logarithmic plot like this, the harmonics fall off at a rate of 1/harmonic#^2, versus the symmetrical square wave where the harmonics fall off at 1/harmonic#.

Now let's try going with a 60% ratio of T$_{Rise}$ to T$_{Total}$. Here's the time domain plot:

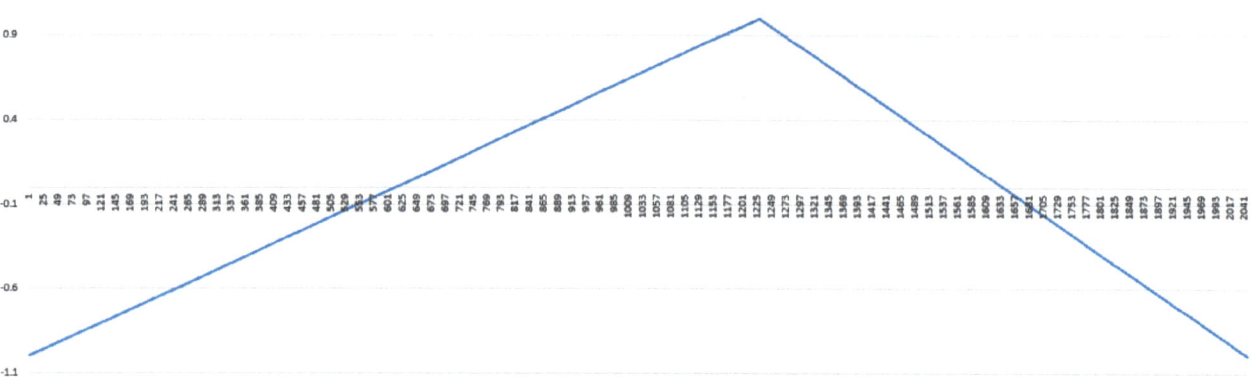

Figure 39 - 60% TR/TT triangle waveform

Note that while the signal is symmetrical in voltage, it is asymmetrical in time—which should, according to my theory, cause **even** harmonics to appear.

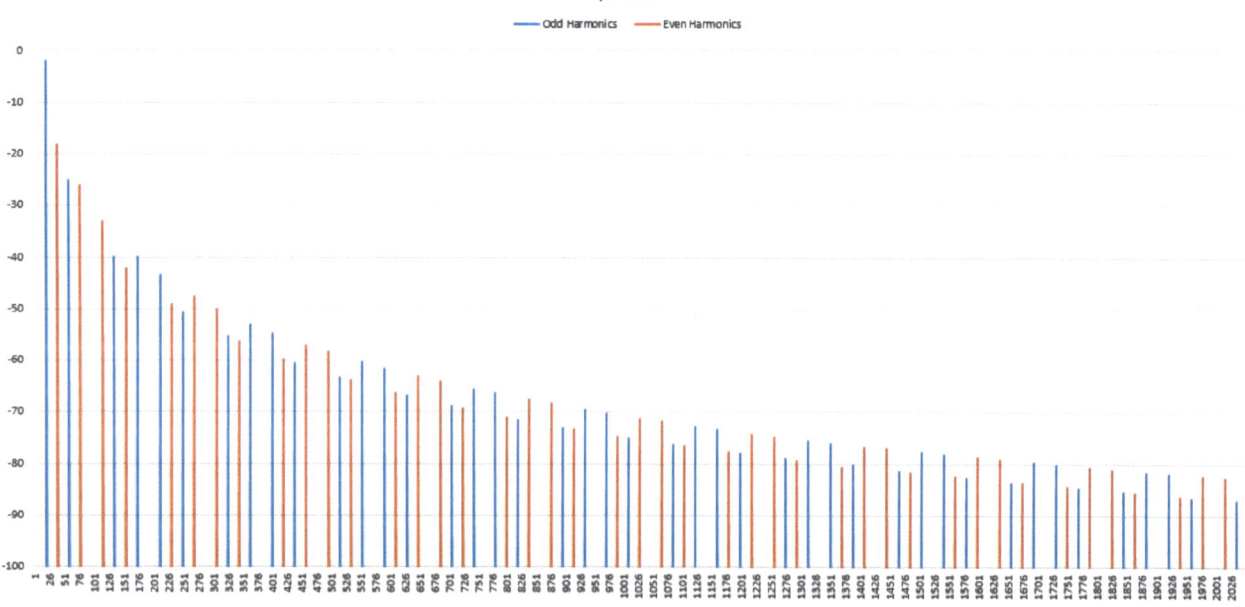

Figure 40 - 60% TR/TT triangle wave spectrum

Suddenly we see the **even** harmonics pop up because of the asymmetry in *time*. The harmonics still fall off in a similar pattern, but because of the sine term in the equation, we now see an interesting [dance of harmonic amplitudes](https://youtu.be/DvOcNDUtA2A)[14]. Once the T$_{Rise}$ to T$_{Total}$ ratio gets away from 50%, the sine function starts creating some lovely patterns in the spectrum. Let's push it up to 70% and see what kind of cool patterns we get:

---

[14] https://youtu.be/DvOcNDUtA2A

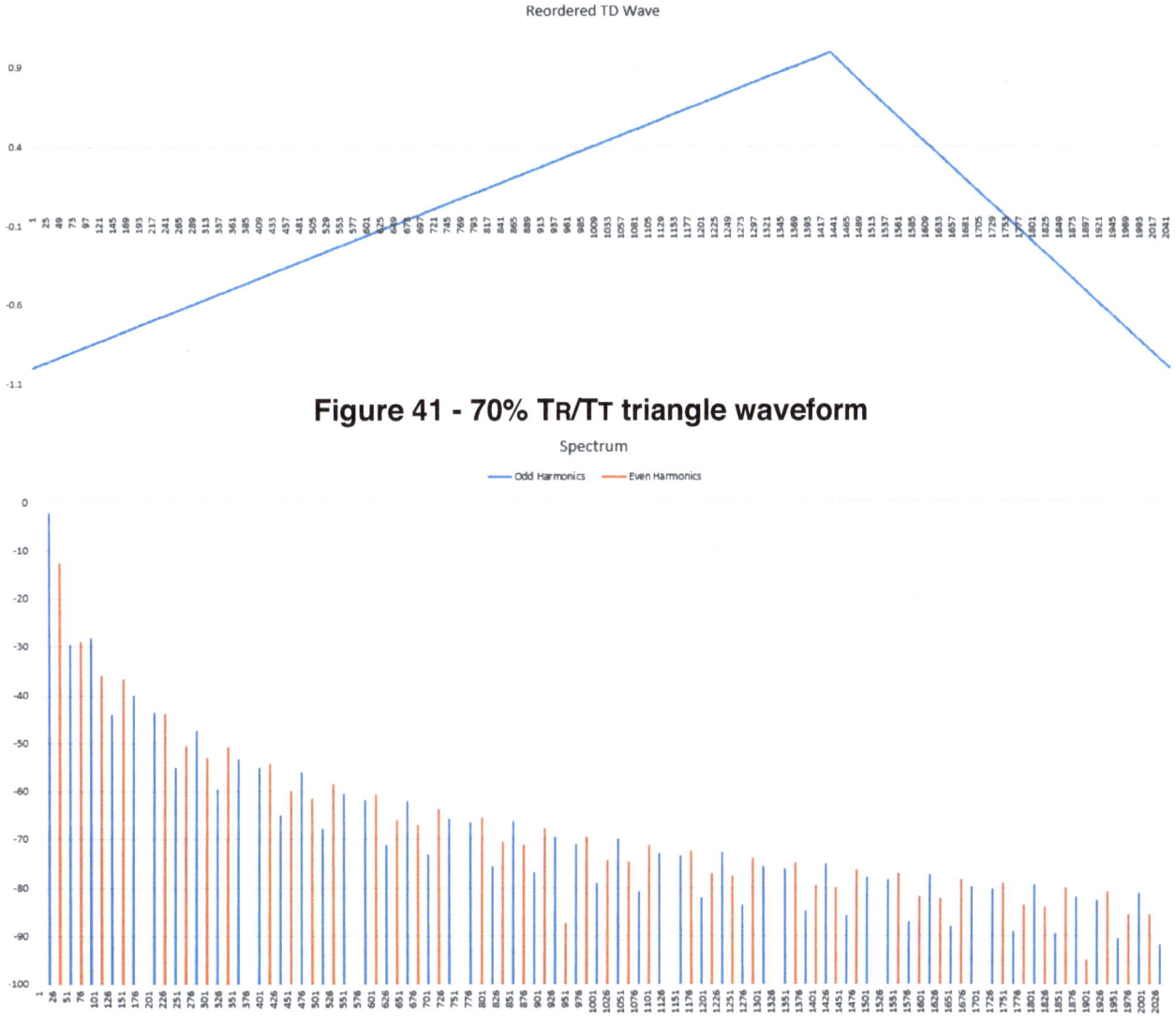

Figure 41 - 70% TR/TT triangle waveform

Figure 42 - 70% TR/TT triangle wave spectrum

More asymmetry, more **even** harmonics. Notice that the second harmonic here is about -12dB, but at a TR/TT ratio of 60%, it was closer to -18dB. It also knocks the third harmonic down to about -30dB; whereas, at a TR/TT ratio of 60%, it was closer to -26dB. Remember though, it's not a good idea to look at these plots harmonic by harmonic because the sine function makes the harmonic amplitudes do a rather interesting dance that defies analysis of individual harmonics. Instead we should be looking for patterns. That is the trap that some engineers of earlier times fell into. They didn't realize that there was a sine function driving the harmonic amplitudes. Because Analog to Digital convertors were slow back then, and because aliasing was so feared (MIT warns against aliasing even today[15]), every capture required an Anti-Aliasing filter. No one looked very high in the spectrum so they could not see that were general rules for harmonics.

---

[15] https://youtu.be/1EI4znkRH0g?t=31m26s

This point gives me an opportunity to restate what Bob Metzler tried to say in his 1992 book, The Audio Measurement Handbook. Whereas he said "non-linearities which are not symmetrical around zero (he meant in zero volts) produce dominantly **even** harmonics," that really isn't the case. As we saw earlier, any non-linearity causes both **odd** and **even** harmonics. When the non-linearity is asymmetrical in either in *time* or *voltage,* the **even** harmonics are no longer suppressed and start to appear. The **odd** harmonics are there all the time for most types of distortion, you cannot get rid of them unless you get rid of all non-linearities. So Mr. Metzler's statement should have been "non-linearities which are asymmetrical in either time or voltage allow **even** harmonics to reappear due to lack of suppression by an equal non-linearity in the opposite voltage polarity or opposite time distortion." This statement is not as pithy as the first, but it is clear that when he wrote that statement he didn't understand that non-linearities in the time dimension have the same effect as non-linearities in the voltage dimension. An unequal duty cycle square wave or unequal rise and fall times in a triangle wave will cause **even** harmonics to re-appear. It is also pretty plain to me that he didn't understand why **even** harmonics didn't appear in symmetrically distorted (in time or voltage) waveforms. The cancellation of **even** harmonics was not a concept he even considered. In fact it's hard to believe that such a thing could exist, that the distortion is there, but can't be detected because of whole cycle cancellation of symmetrical distortion. But as I mentioned in chapter 1, you cannot hear a half cycle. If you try to rectify a waveform so that only one half cycle is retained and the other is eliminated, you have unavoidably introduced far more distortion than you might have been looking for in the original wave. Very soon though, I will show you a waveform that confirms this statement.

Here is the crucial fact that you must finally realize. An FFT does not just describe reality, it *is* reality! Your ears agree with the FFT: so do other mechanical structures. Don't imagine that an FFT is just for seeing what something is doing. The FFT describes reality in another dimension, another domain–the frequency domain where time is irrelevant and frequencies rule.

I want you to see how this goes as we keep increasing the TR/TT ratio (or decreasing the TF/TT ratio if you look at the opposite side). On to a 90% ratio, time domain first, then frequency domain:

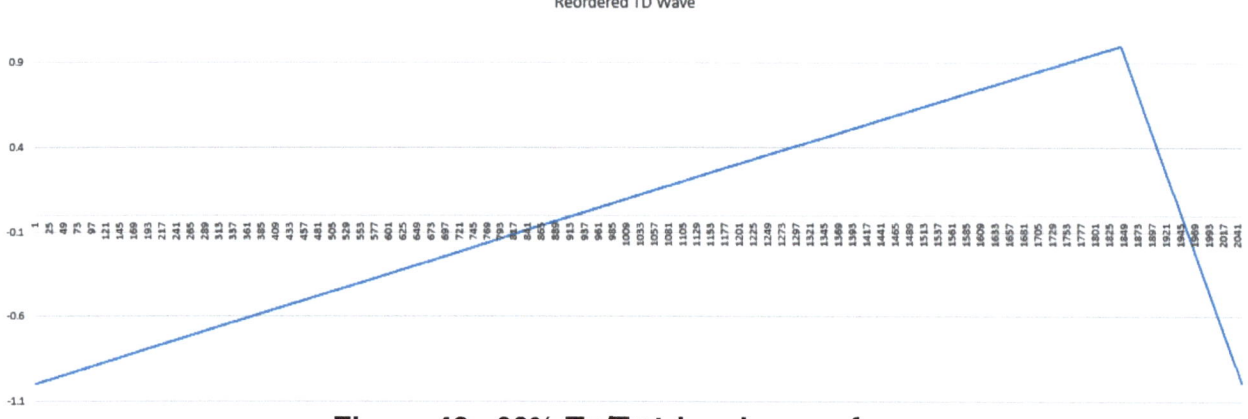

Figure 43 - 90% TR/TT triangle waveform

Figure 44 - 90% Tʀ/Tᴛ triangle wave spectrum

Again those lovely patterns. The pattern of humps punctuated by notches is due to the sine function in our formula, but realize that these plots are not made by the formula, they are made by performing an FFT on the waveform above in figure 43. The formula simply predicts and explains why the spectrum looks the way it does.

Earlier I mentioned the sawtooth wave which was used for decades for oscilloscopes and television horizontal scanning. The 99% Tʀ/Tᴛ ratio wave below is a close approximation of some of the early sawtooth waves. Charging a capacitor through a high impedance, then discharging it through a low impedance to get a fast retrace time. Of course, the retrace time has to be fast (1% or less if possible) to avoid missing any crucial signal details on an oscilloscope; and, during that time the electron beam had to be *blanked* to prevent the user from seeing this faint trace across the screen in the opposite direction of the main trace. The faster the better, and faster, of course, means more harmonic content persisting much higher in the spectrum. As it turned out, in those days before surface-mount components and printed circuit boards, it was pretty hard to get the bandwidth to get a fast retrace time, and fast in time is high in frequency.

If you want to see what a slow, un-blanked retrace looks like, watch the Star Trek episode Mudd's Women[16]. In the scene where the "computer" is analyzing Harry Mudd's voice for signs of deceit, they used an older oscilloscope to display the voice of the computer. Watch closely and you can see the retrace going right to left across the screen.

---

[16] http://www.hulu.com/watch/283804

Now for the plot of a 99% TR/TT ratio wave:

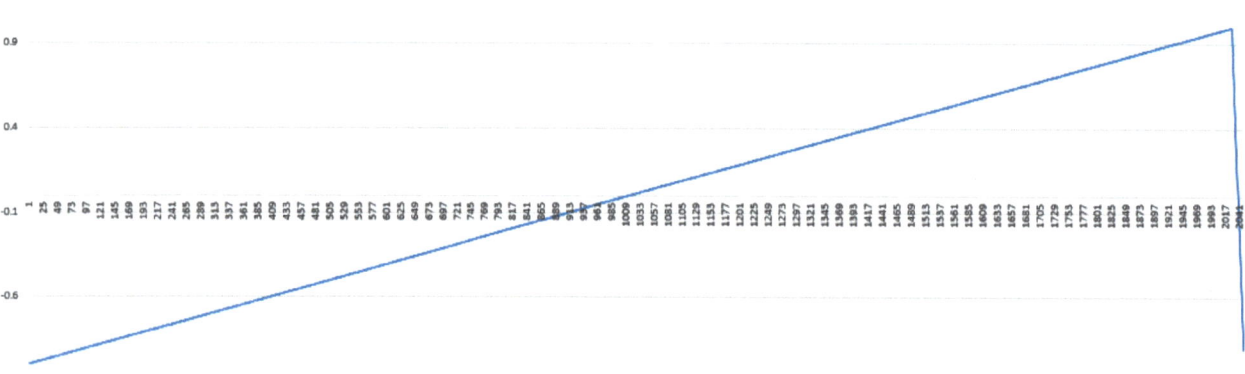

Figure 45 - 99% TR/TT triangle waveform

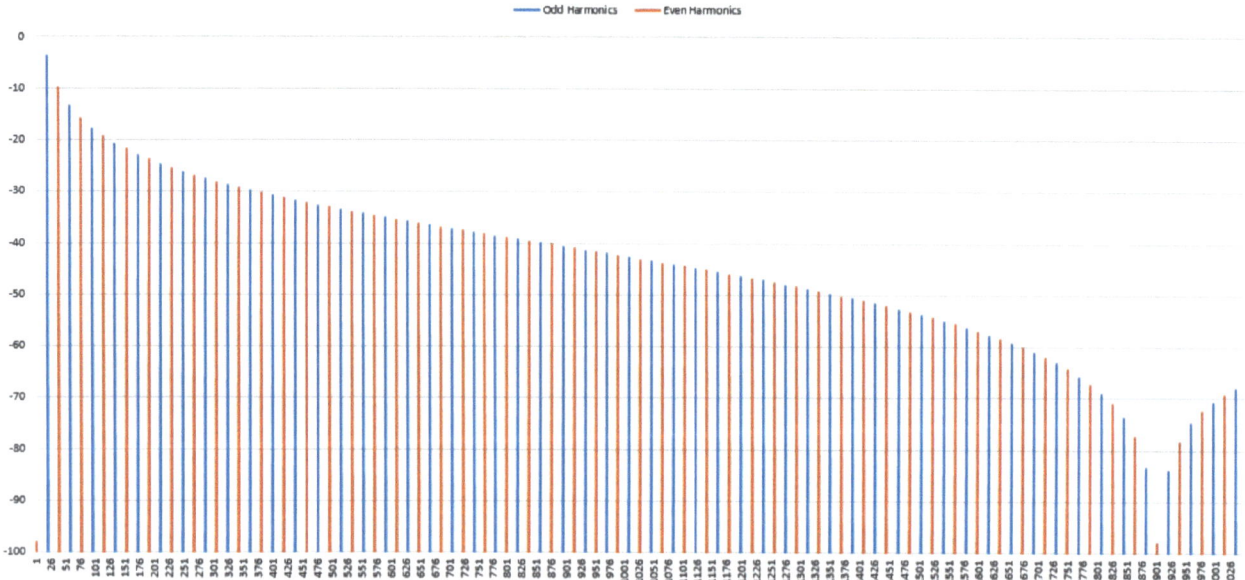

Figure 46 - 99% TR/TT triangle wave spectrum

Notice how the notch is heading off to the right, about to disappear off the spectrum. That notch is the last remnant of the 10 notches you saw back in figure 44. As the TR/TT ratio goes even higher, the edge gets faster requiring much higher frequencies. The sine function pushes those notches higher and higher in frequency until they disappear altogether.

The next, and last pair of plots in this chapter will confirm this fact:

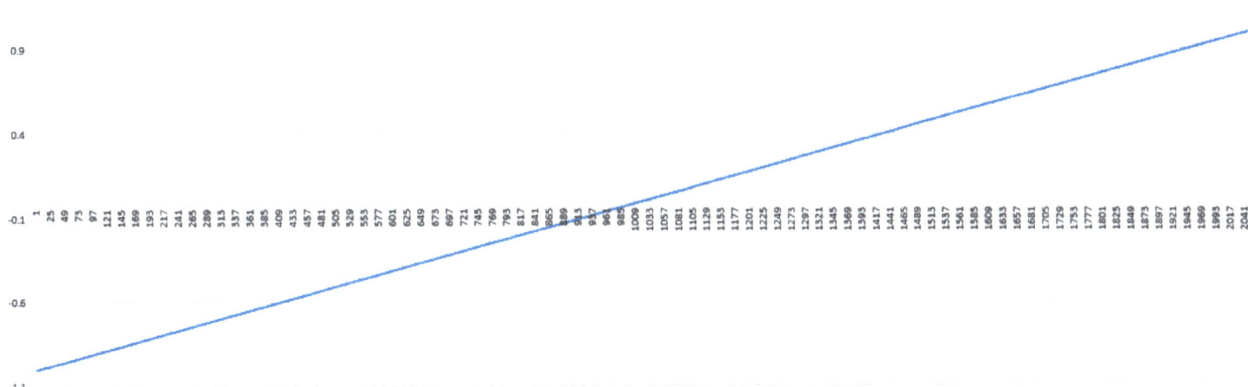

Figure 47 - 100% TR/TT triangle waveform

At a TR to TF ratio of 100% there is no ramping down, only ramping up, very similar to our single impulse rectangular wave. This infinitely fast transition at the end of the ramp is hard to do in a real world circuit, just like infinitely fast edges of a square wave are hard to do with limited bandwidth. Remember that the world is a low pass filter; most engineers spend a lot of time fighting that reality to get their edge rates up, and for fast edges, you need very high bandwidth. It used to be said that you need only worry about the first 11 harmonics to get a decent square wave or sawtooth wave. That's not even close to the truth, as you can see here. Those high harmonics are responsible for those very fast edges.

Now for the spectrum. Prepare to be amazed.

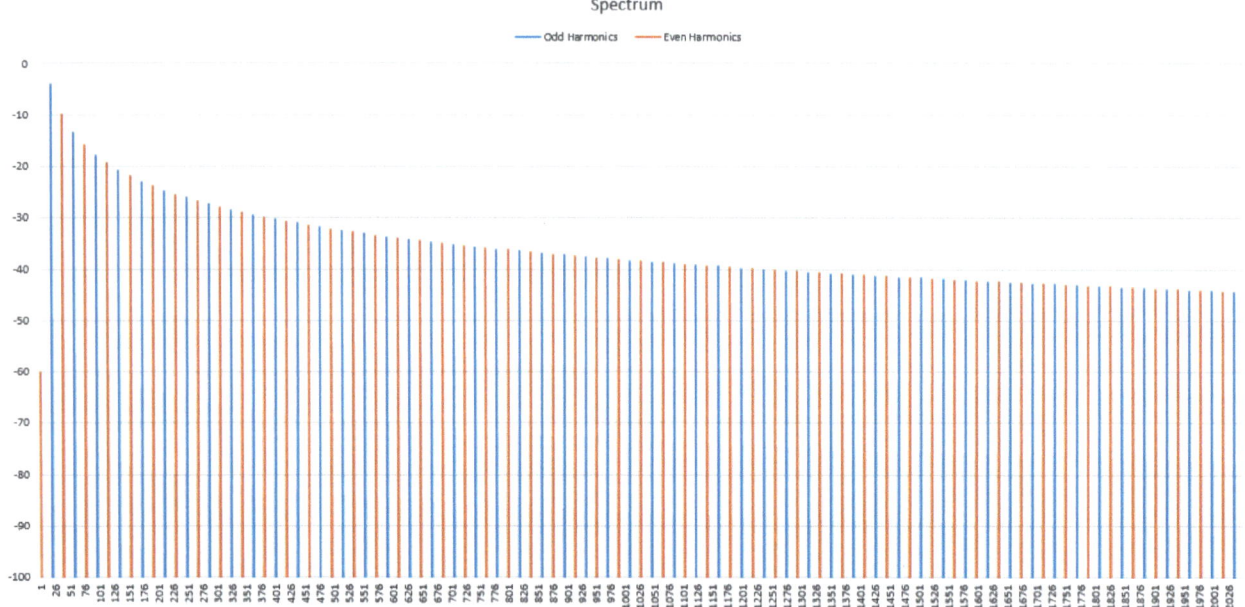

Figure 48 - 100% TR/TT triangle wave spectrum

Looking at the spectrum you might notice that it has a familiar appearance. Of course it's filled with **odd** and **even** harmonics; it's the most asymmetrical triangle/sawtooth wave you can get, so obviously it's going to have both **odd** and **even** harmonics. But the rate at which the harmonics decay might seem familiar. That's because you saw a spectrum similar to this back in chapter 1 in figure 11. Rather than make you go back there, let me show you:

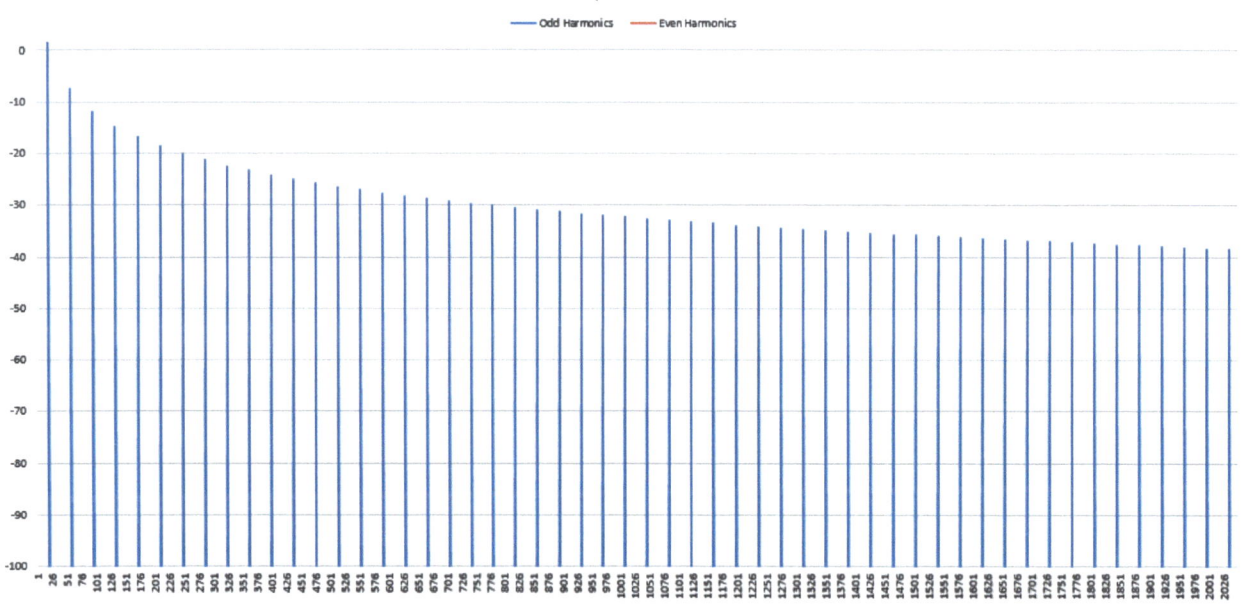

Figure 11 redux - 50% duty cycle square wave spectrum

Again, because these plots are in dB it might be hard to recognize, but with a 100% TR/TT ratio, the harmonics fall off at 1/harmonic#, just like a 50% duty cycle square wave did! As I told you back in chapter 1, the harmonics of a 50% duty cycle square wave fall off such that the third harmonic is one third the amplitude of the fundamental, the fifth harmonic is one fifth the amplitude of the fundamental and so on. Surprisingly, a 100% TR/TT ratio triangle wave also falls off in this order, but it also has **even** harmonics so that the second harmonic is one half the amplitude of the fundamental, the fourth is one quarter, the sixth is one sixth, etc, which changes the shape of the wave dramatically.

Below is a table showing the differences and similarities between a 50% duty cycle square wave and a 100% TR/TT ratio triangle wave. Note that I color coded it the same way as my spectra, blue for **odd** harmonics, red for **even** harmonics. Note that the amplitudes for this wave are relative to a normalized one volt fundamental because, as you might have noticed above, the fundamental and every harmonic of my ramp spectrum is down by *exactly* 6dB from the square wave spectrum.

Harmonic	50% DC Square wave	100% TR/TT Triangle wave
1	1V	1V
2	0V	0.5V
3	0.3333V	0.3333V
4	0V	0.25V
5	0.2V	0.2V
6	0V	0.1666V
7	0.1428V	0.1428V
8	0V	0.125V
9	0.1111V	0.1111V

It is quite an astonishing fact that a highly asymmetrical sawtooth wave has exactly the same spectral signature of a 50% duty cycle square wave, but with the addition of **even** harmonics in proportions that match the **odd** harmonic proportions. Why do you think that is?

What causes those high frequency harmonics in a square wave? The very fast edges, of course. A ramp has one very slow edge, and one very fast edge. Since there is only *one* very fast edge in a ramp, unlike a square wave, there is nothing to counteract the **even** harmonics; hence, the **even** harmonics pop up in exactly the same ratio as the **odd** harmonics in a square wave. In essence, a ramp wave is a square wave with one half missing, giving you the same harmonic signature as a square wave, but with the **even** harmonics which are no longer canceled out by a complementary edge in the other half of the wave. The case of the ramp **proves** that **even** harmonics are canceled out by symmetry and that the symmetry is not just in *voltage*, but also in *time*, meaning that harmonics are caused by non-sinusoidal components in proportion to their area (voltage*time).

Not just that, but once again we prove that a ramp is half of a square wave because the fundamental and each harmonic of the ramp spectrum is exactly 6dB lower than the **odd** harmonics in the square wave spectrum. One fast edge, half the energy of a square wave that has two fast edges. The square wave has twice the energy, but without the **even** harmonics which aren't there because they are canceling each other out.

Finally, how would a triangle wave come about in nature? Remember that if you overdrive an amplifier you get a square wave as in this video where I convert a sine wave into a square wave[17]. If you were to drive a high frequency sine wave into an analog circuit with a limited slew rate, the result would be a triangle wave. Because sine waves move very fast at the zero crossing, if the input frequency *dv/dt rate* exceeds the slew rate, the circuit would not be able to keep up and the output would be a low amplitude triangle wave that would ramp up and down at the maximum slew rate of the device. We already know that triangle waves produce a plethora of harmonics, and if the slew rate on one slope differs from the slew rate of the opposite slope, you can bet there will plenty of **even** harmonics mixed in with the ever present **odd** harmonics. Slew rate limitations can seriously impact sinusoid purity by causing the output to deviate from the perfect sine wave in the *time* dimension which as we have seen, will cause harmonics to appear.

It seems fitting that a square wave would be the result of overdriving beyond the *voltage* capabilities of the circuit, and a triangle wave would be the result of overdriving beyond the *speed* capabilities of the circuit.

---

[17] http://youtu.be/yxc-GCmYKXE

Chapter 3
Distortion Signatures - Clipping

We've studied the harmonic signatures of square waves and triangle waves and their derivatives, but most of the time engineers are concerned with unwanted harmonics that present themselves when a circuit is stimulated with a sine wave. That is why THD is measured with a very specific sinusoid applied. The reason is pretty simple. Non-linearities do not *announce* themselves to the world. They must be teased out of the circuit through some form of stimulation. In a standard linearity test, that is done with a ramp waveform which reads out the non-linearities directly, but in THD testing we use a sine wave. Why?

When a sine wave is applied to a perfectly linear transfer function, a sine wave comes out with *no additional spectral components*. Any non-linearities disturb the applied sine wave and hit those speed bumps we call non-linearities, creating unwanted spectral components.

Try this fun little experiment. Find a disposable paper or plastic cup and drop a marble in it. Now hold the cup in your hand and wiggle it around so that the marble is spinning around the inside of the cup. If your grip on the cup is loose enough, your fingers will not distort the path of the marble and it will travel in a perfect circle around the inner circumference of the cup. Now tighten your grip, distorting the sides of the cup while the marble orbits inside. Suddenly the distortions will disturb the orbit of the marble and cause it to follow a non-circular path. You might even be able to get it to bounce around inside the cup as you try to keep the marble orbiting. With a tight enough grip, the distortions on the side of the cup will make it impossible for the marble to do anything other than bounce around inside the cup. Loosen your grip, and if you haven't permanently damaged it, the marble will continue its perfectly circular orbit. You have simulated a variable distortion mechanism in an otherwise perfectly linear transfer function, the inside of a perfectly formed cup.

For historical reasons THD testing is the technique that won out over a normal ramp based linearity test. Back in the past, Digital Signal Processing was unheard of. Engineers knew how to make analog sine wave generators, and they could build RLC analog filters. THD could be measured by sourcing a sine wave into a circuit and using a *notch* filter to take out the fundamental; then, the RMS voltage of anything left behind was assumed to be harmonic energy. A second filter, a *lowpass* filter, might be applied to keep the number of harmonics measured limited to some arbitrary value. Before "high fidelity," or HiFi for short, measuring THD over nine harmonics probably seemed pretty generous.

But now that we can digitize almost any frequency, either directly or indirectly (by using a down-convertor for very high frequencies) there is no excuse for not digitizing the waveform for a closer examination. Thanks to the invention of the FFT by Tukey and Cooley, we can now look at any tone we want and find exactly which tones are in a test waveform, and figure out which harmonics are created by which distortion anomaly.

Now that you have a good feel for where harmonics come from and the kinds of things that can create harmonics, it's time to look at various hardware faults and connect the harmonic signatures to them. The overall goal of this book is for you to be able to look at the output of a spectrum analyzer and say "Oh, I see some crossover distortion there" or "I see that the device is clipping one side but not the other." Wouldn't it be nice to be able to do that? Remember, there is a reason we use spectrum analyzers, it's sometimes very, *very* difficult to see device faults in time domain: but, if you pay attention to the contents of this book, you should be able to figure out where to look when you see a problem in a spectrum. The alternative is to act like a shade-tree mechanic and swap out components or tweak resistors until the problem goes away. This approach has been used for years, and it takes much longer than just analyzing the spectrum and pinpointing the problem in a few minutes.

Of course to do that, you will have to remember the odd/even patterns and note the way the harmonics build and decay. Sine functions can make some pretty patterns, and you have to remember that the *humps* and *notches* you see in the spectrum are real artifacts of the sine function that we saw in both square waves and triangle waves. It has nothing to do with frequency response or anything else. I can't give you the exact formula for every kind of distortion that you might see, (I will eventually give you a formula for simple distortions), but all distortion causes harmonic products with the same factors we saw in square waves and triangle waves. Remember that for square waves the formula was;

$$\texttt{VHarmonic = abs(sin(}\pi\texttt{*Harmonic\#*Duty\_Cycle))/Harmonic\#}$$

And for triangle waves it was;

$$\texttt{VHarmonic = (abs(sin(}\pi\texttt{*Harmonic\#*TRise/TTotal))/Harmonic\#\^2)/}$$

$$\texttt{sin(}\pi\texttt{*TRise/TTotal)}$$

The formulas, and thus the behavior of every other kind of distortion will be similar. You will learn that the *phase angle* where the sinusoid hits the distortion determines what the harmonic distortion pattern will look like. If you pay attention to this pattern, it will tell you where in the device's transfer function the distortion lies.

Asymmetrical Clipping Distortion

When a sine wave clips at either the top or bottom, it's pretty easy to see in the spectrum. The first thing to remember is that both odd and even harmonics will be present, unless the clipping is symmetrical; then, only odd harmonics will be present. Of course, it would have to be very, very symmetrical, any asymmetry will cause even harmonics to pop up. That's what I was trying to prove in the video I made for Applicos, diddling the dc offset to try to attain pure symmetry in clipping, ridding myself of those even harmonics.

Let's start with a 4% clipping on the top of a pure sine wave. By this I mean that I will corrupt 400 samples in a 10,000 sample transfer function that will cause clipping of the sinusoid at 66.8 degrees, then hold that level through 90 degrees and all the way through to 113.2 degrees. First the time domain wave, then the spectrum:

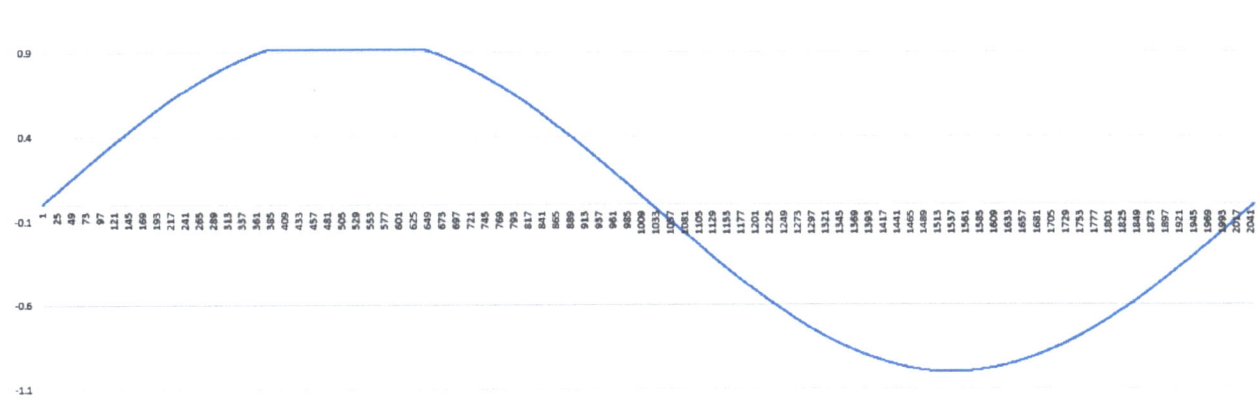

Figure 49 - 4% upper peak clipping distortion, time domain

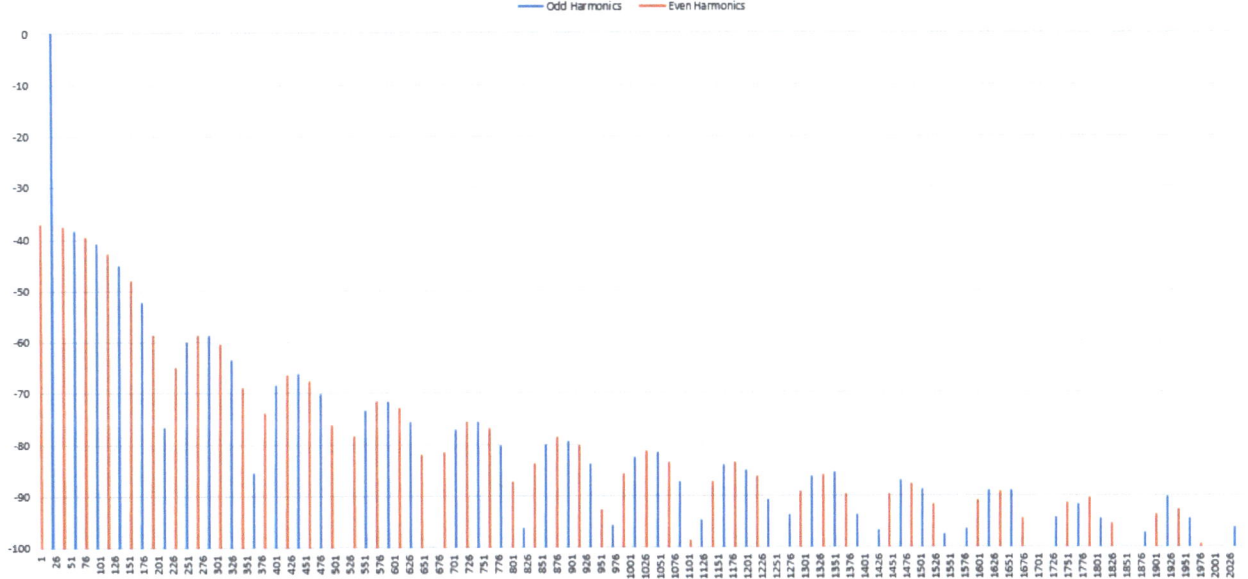

Figure 50 - 4% upper peak clipping distortion spectrum

You should have expected to see both **odd** and **even** harmonics, and if you didn't, you haven't been paying attention.

The THD for this wave is -32.16dB for the first 8 harmonics and almost the same, -32.099dB for the first 50 harmonics. The reason for the similarity between the two numbers is obvious. Clipping creates a harmonic signature that starts with high harmonic amplitudes low in the spectrum; then the harmonic amplitudes fall off very quickly. This is what most engineers expect, treating harmonic distortion as though it were subject to some low pass filter. But that is absolutely not the case, as you have seen so far. The extent of the roll-off is not a function of any filter; it's a function of the very slow curve of the peak being removed from the wave. Remember what I said earlier; harmonics are created when area is added to, or removed from the sinusoid. When you remove the very slow peak, the harmonic curve mirrors the spectral content of that feature[18]. The peak of a sinusoid is a

---

[18] https://www.linkedin.com/pulse/harmonic-signatures-ii-dan-bullard/

very slow creature, hence the distortion energy is concentrated in the low frequency harmonics and less in the higher frequencies.

For this wave the **even** THD is -34.52dB for 50 harmonics, the **odd** THD is almost the same at -35.78dB, again for 50 harmonics. When I separate the **odd** and **even** THD, it's always for 50 harmonics, 25 **odd** harmonics and 25 **even** harmonics. The fact that these two numbers are so close tells you that there is no symmetry in the clipping. As you saw back in chapter 1, distribution of harmonic energy between **odd** and **even** harmonics is a clue to the symmetry of the distortion. If there was any clipping on the bottom side as well, it would cancel out some of the **even** harmonic energy making **odd** THD higher than **even** THD. Sometimes Bob Metzler's statements on harmonics almost seem correct, but they're not.

Note that the first notch bottoms out at the 11th harmonic. Note also that there are 13 humps in this spectral plot. While this varies with sampling frequency and number of samples, be aware that I will not change either in this book so you can use the number of humps and notches as a clue to how each distortion highlighted changes the spectrum from example to example.

So, why are there 13 humps? It's all based on the *angle* where the distortion occurred. Just as changing the duty cycle of our square wave caused the number of humps to change in our spectrum, the phase angle where the distortion intersects the sinusoid determines how many humps occur in our spectrum. Think about what duty cycle means to a square wave. A 50% duty cycle square wave makes the transition from **high** to **low** at 180 degrees, so there is perfect symmetry in both voltage and time (and therefore no **even** harmonics). If the duty cycle is other than 50%, the transition intersects the implied underlying sinusoid at some angle other than 180 degrees. In our formula for the harmonic amplitudes, we used duty cycle, as defined by **high time** divided by **total time** to control the humpy-ness of our spectrum, as in:

`VHarmonic = abs(sin(π*Harmonic#*High_Time/Total_Time))/Harmonic#`

But wouldn't it be just as valid to use the location of the high to low transition in degrees relative to 360 degrees?

`VHarmonic = abs(sin(π*Harmonic#*H2L_degrees/360))/Harmonic#`

It would be the same, right? Don't worry that I am mixing degrees and radians. H2L_degrees/360 is just a ratio. If you feel squeamish about this, use phase in radians over 2π. I prefer to think in terms of degrees.

Normally we don't think about duty cycle this way, but expressing duty cycle in this manner is just as valid as high time over total time. And guess what? The results are exactly the same! So, it's not so much the duty cycle that changes the signature of a square wave's spectrum (or a triangle wave's for that matter), but the *phase angle* at which the transition occurs relative to the expected 180 degrees of a normal wave. Don't believe me? Let me prove it.

Remember in the above example the distortion (clipping) starts at 66.8 degrees as the sinusoid makes its way to the peak at 90 degrees and continues on to 113.2 degrees where the normal sinusoidal shape resumes. What if I change the test and do not clip, but instead place a small distortion at 66.8 degrees, which naturally replicates at 113.2 degrees as the sinusoid passes through the same place in the transfer function? Notice that this distortion happens at exactly the same place on the sine wave as our peak distortion in figure 49.

The only difference is that it does not hold the peak at that level, the wave continues on as normal after this small distortion.

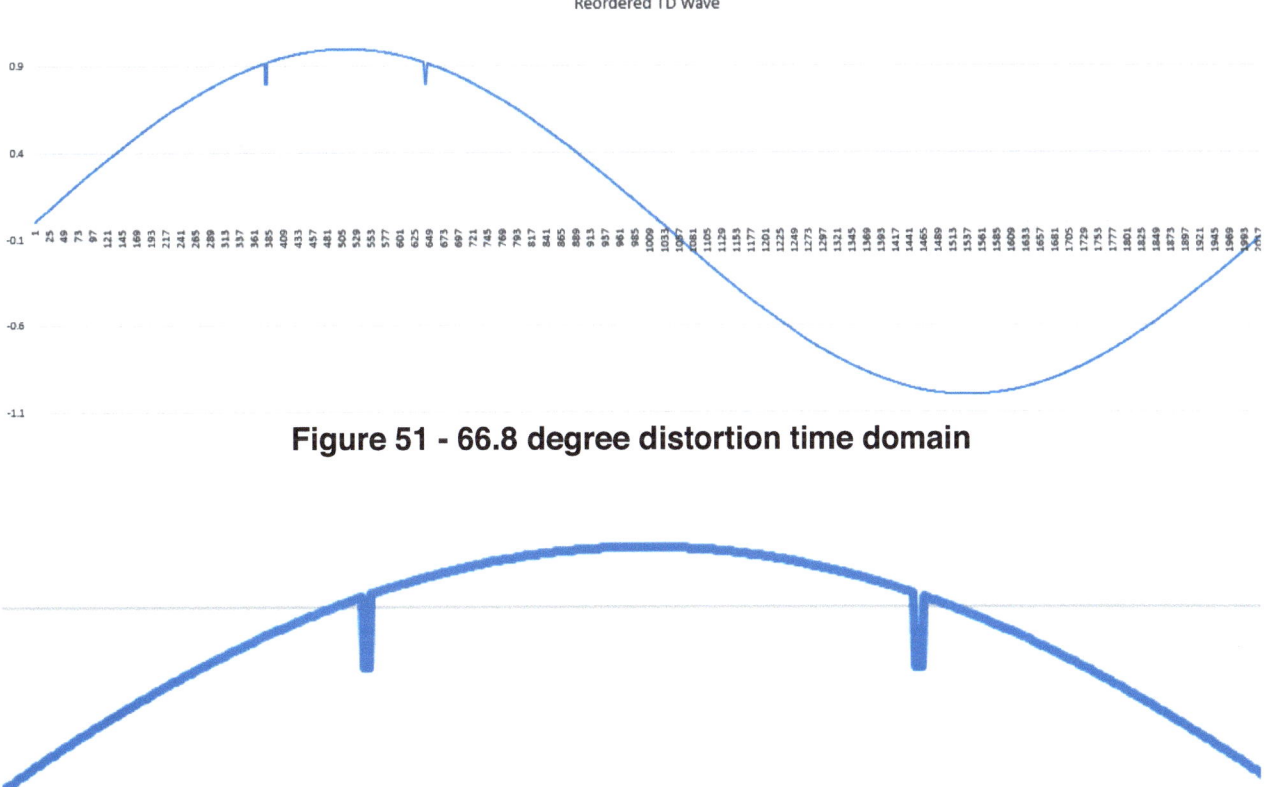

Figure 51 - 66.8 degree distortion time domain

Figure 52 - 66.8 degree distortion time domain closeup

Notice that this spectrum has 13 humps, just like the clipping distortion in figure 50 that had 13 humps.

Figure 53 - 66.8 degree distortion spectrum

Also, the second notch is at the 12th harmonic, not the 11th, but close to what we saw in figure 50. You will also notice the 107th harmonic popping up in blue at the very right edge of the spectrum in both figure 50 and 53. These similarities are not a coincidence. The phase angle of the distortion is what controls the number of humps and the location of the first notch, and subsequent notches. Notice also that compared to the clipping spectrum, this one has an earlier notch at the 4th harmonic, but the clipping spectrum has a lot of low frequency harmonics which is due to the fact that we chopped the entire top off of the sinusoid. This adds a lot of low frequency energy, so we don't get the first notch at the 4th harmonic as we did in the very small distortion spectrum.

You will always see this behavior with clipping distortion. The first hump will seem wider than all the others because the surplus of low frequency energy fills in the first notch with a very robust first hump. Then because of the lack of fast edges in the distorted wave, the amplitudes of the harmonics roll off rather quickly, emulating the roll-off seen in a Butterworth filter and tricking us into thinking that there is some inherent low pass filter involved in the harmonic distortion process. There is not, and now you know the real reason for this appearance that has fooled hundreds of thousands of scientists, engineers and professors in the past and present.

Total Distortion Energy

In order to explain what is going on with harmonic distortion, I'm going to introduce a parameter I call Total Distortion Energy, or TDE, measured in Volt-seconds or V-s. This parameter describes how much total *area* of a transfer function is disturbed by a distortion anomaly.

In the simulations I show here, I used Excel to implement a 10,000 point transfer function from -1 to +1. Since this transfer function represents a rectangle, I can calculate the total area of the transfer function. I would like to propose that we scale the *width* of the transfer function by 1ms total. Traditionally THD is tested with a 1KHz tone, which has a period of 1 millisecond. If we scale TDE such that we assume that the entire transfer function will be swept by a 1KHz sine wave in one cycle, then we can attribute a value of 1 millisecond of time to the width of the entire transfer function. My transfer function is 2 volts high and 1 millisecond wide. Because the area of a rectangle is h*w you can see that, in total, my transfer function contains the potential of 2 volts times 1ms or 2 milliVolt-Seconds (2mV-s) of area (energy).

When I clip the top 400 points at 0.919991999, which is how I simulated peak clipping in figure 49, I end up with a chunk stolen from my transfer function in the form of a triangle, as you can see below.

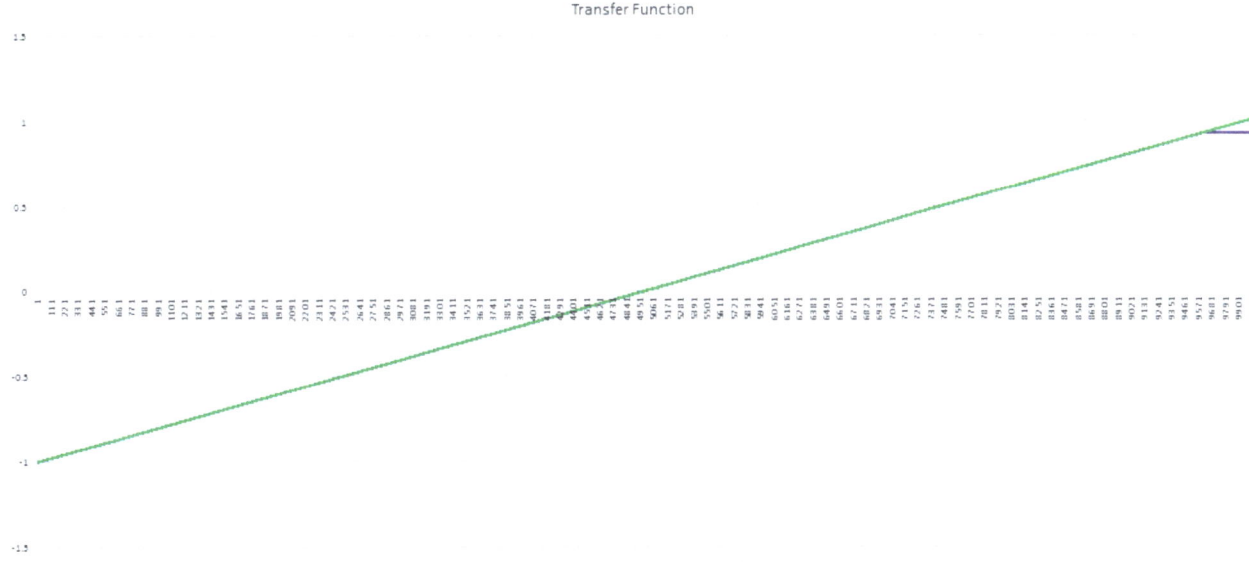

Figure 54 - 4% distorted transfer function distortion

Figure 55 - 4% Transfer function distortion closeup

We can calculate how much area this triangle has. This triangle has a height of (1.0-0.919991999) or 80mV and a base of 400 points, which, in this case, represents an area of 0.5*80mV*400/10000*1ms or 1.6uV-s of TDE. This represents a small, but not insignificant area stolen from my transfer function. I call this a 4% distortion, because I have compromised 4% of the transfer function curve. That does not translate directly into energy however, because I can do all kinds of crazy distortions that can't be described in anything but total area, which is why I devised the TDE specification. This particular distortion has 1.6uV-s of TDE, which translates into harmonic energy which resulted in a THD of about -32dB, as we saw above.

Now, to show what changes when the amount of distortion changes, let's cut the distortion down to 2%, half of what we saw above. In this example, the Total Distortion Energy is 0.4uV-s of TDE, one quarter of our previous example. Why one quarter instead of one half? Because we are clamping the last 200 samples of the transfer function to 0.959999V. If we do the math, we find that not only has the base of the triangle shrunk by a factor of two, but the height has shrunk by a factor of two as well. So now the TDE is 0.5*40mV*(200/10000)*1ms or 0.4uV-s of TDE.

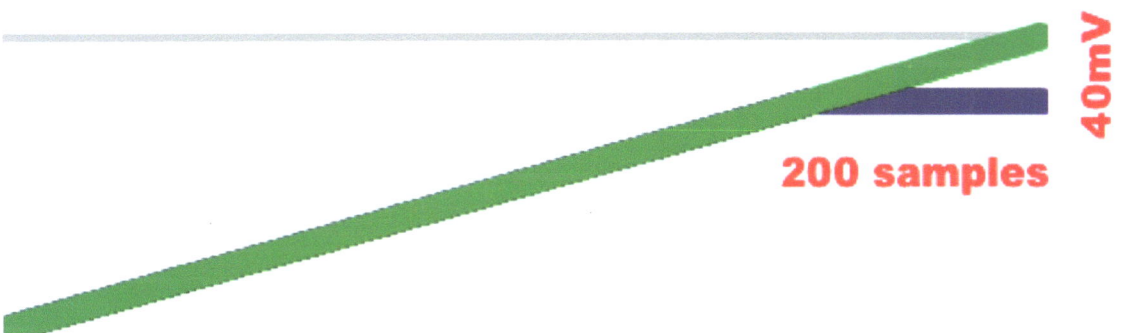

Figure 56 - 2%, 0.4uV-s TDE upper peak clipping distortion

We apply the full scale sinusoid to the transfer function, something we can't do in a real device because of the danger of clipping, and now we look at the plots:

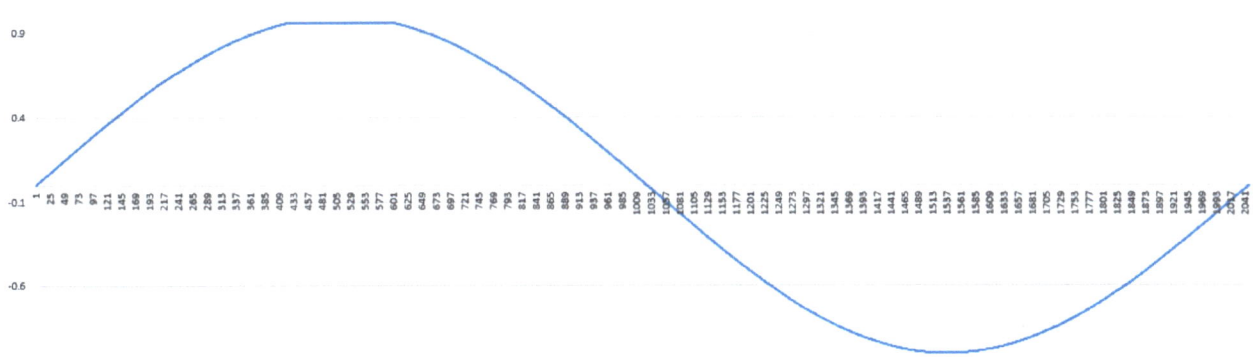

Figure 57 - 2%, 0.4uV-s TDE upper peak clipping distortion, time domain

Now we perform an FFT and check the result:

Figure 58 - 2%, 0.4uV-s TDE upper peak clipping distortion spectrum

Notice that the number of humps is fewer and they are more spread out. There are 9 humps, and the first notch happens at the 16th harmonic. Why did this happen?

Because the clipping distortion started at 73.75 degrees and continued on to 106.25 degrees. What do you think would happen if I repeated the previous experiment and placed a tiny distortion in the transfer function that would intersect my sinusoid at 73.75 degrees and 106.25 degrees?

Figure 59 - 73.75 degree distortion spectrum

Nine humps, just like what we saw in figure 58. It's not a coincidence: the number of humps is due to the *angle* where the distortion intersects the sinusoid. Again, the longevity of the harmonic amplitudes going out to the right is due to the fast edges in my very tiny distortion. The rapid drop off in harmonic energy in the clipping example is due to the fact that there are no fast edges at all. In fact, there is a plethora of low frequency energy that fills in the first notch at the 6th harmonic because clipping removes the top of the wave. The nice slow cresting of a sine wave has a very low frequency spectrum. Remember that harmonics can come from energy added to, or energy *removed from,* the implied underlying sine wave, just like we saw in figures 26-29.

The THD from our spectrum in figure 58 is lower than in the previous example: -39.5dB for 8 harmonics, -39.16dB for 50 harmonics, -41.78dB for **even** harmonics, -42.6dB for **odd** harmonics. Note that a TDE of 0.4uV-s resulted in a THD of -39dB versus the previous waveform, with 1.6uV-s of peak distortion, that had a THD of -32dB. That's about a 7dB difference for a factor of four decrease in distortion energy. That might not make sense, you might have expected something more like a 12dB drop in THD (a factor of 4 drop in THD for a factor of 4 drop in distortion area) but that is not how it works, at least not at the peaks. Later I'll explain this fully.

At the top and bottom of the transfer function relatively small distortions create lots of THD, but *changes* in the distortion area (energy) do not result in comparable changes in THD. That is one of the problems with THD: it does not accurately reflect the damage done to the purity of the signal in a linear fashion. One of the reasons why harmonic distortion is so misunderstood is that these strange effects of non-linear response happen at the edges of the transfer function *where no man has gone before* to paraphrase Star Trek. No one in their right mind would test THD to the very edges of a device's transfer function like this, so virtually no engineer has ever tried to figure out what the response of a sinusoid applied to the very edge of the transfer function might be. But by doing theoretical studies with mathematical models of a real device, we can test the very edges of the model and deduce the mysteries that have been hidden by fear and incorrect thinking—or, as I put it in my Applicos video, we can now learn the very depths of this (or any) device.

Now let's reduce the clipping even more, down to 1% with a TDE of 0.1uV-s. The clipping begins at 78.5 degrees and goes all the way to 101.5 degrees. If previous examples are an indicator, the number of humps will go down as the distortion gets closer to 90 degrees.

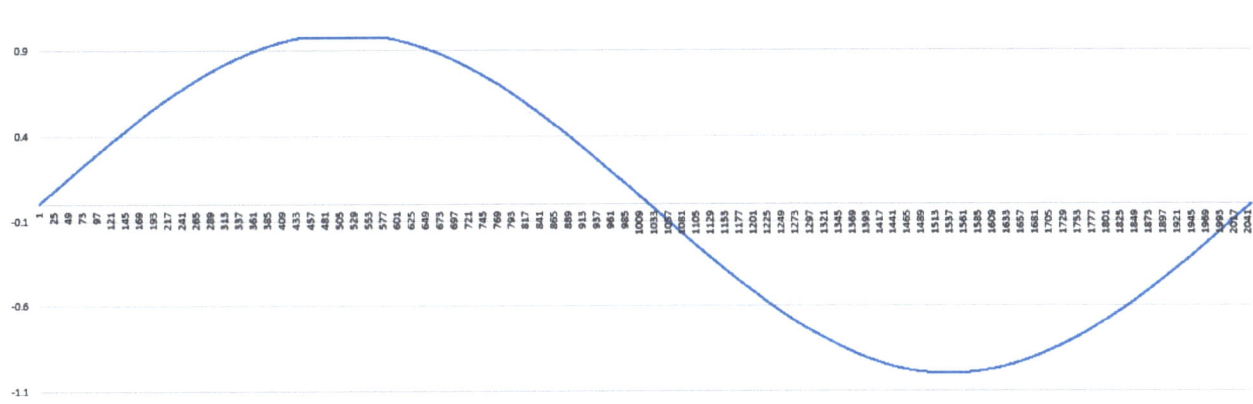

Figure 60 - 1%, 0.1uV-s TDE upper peak clipping distortion, time domain

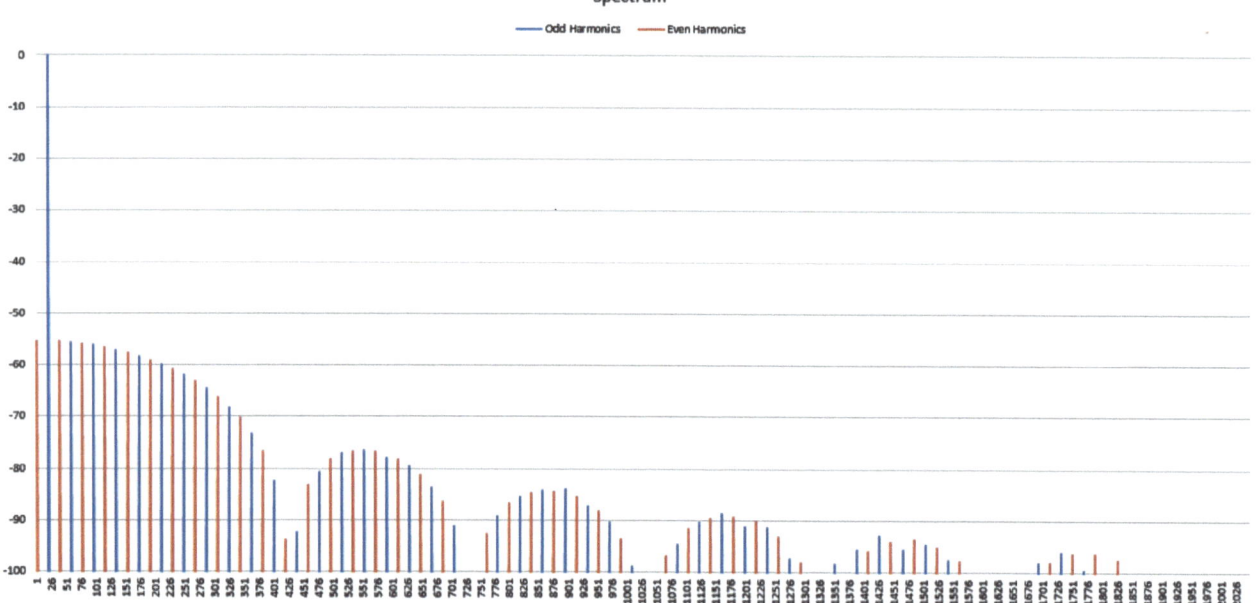

Figure 61 - 1%, 0.1uV-s TDE upper peak clipping distortion spectrum

Six humps, just as we expected. The THD is now -47.46dB for the first 8 harmonics and -46.34dB for the first 50 harmonics. **Even** THD is now -49.1dB; **odd** THD is -49.63dB. The fact that the TDE went down by a factor of 4, yet the THD only went down by a little over a factor of two, -46dB vs -39dB, a difference of about 7dB, means that sine waves are less sensitive to changes in distortion energy at the peaks.

To recap, let's look at the results we got for this series of asymmetrical clipping:

Distortion	TDE	THD	Humps	Degrees
4%	1.6uV-s	-32.1dB	13	66.8
2%	0.4uV-s	-39.16dB	9	73.75
1%	0.1uV-s	-46.34dB	6	78.5

Each halving of distortion samples reduces the TDE by a factor of four. That should be obvious, because as I reduce the number of points corrupted along the transfer function, the triangle formed by distortion of the transfer function gets smaller, and thus, the area goes down by the square. The fact that the THD goes down by 7dB for each halving of the samples distorted in the transfer function is unexpected, because each halving of the number of distortion samples decreases the area by a factor of four, which should reduce the THD by 12dB. But there is a simple explanation for what is going on here.

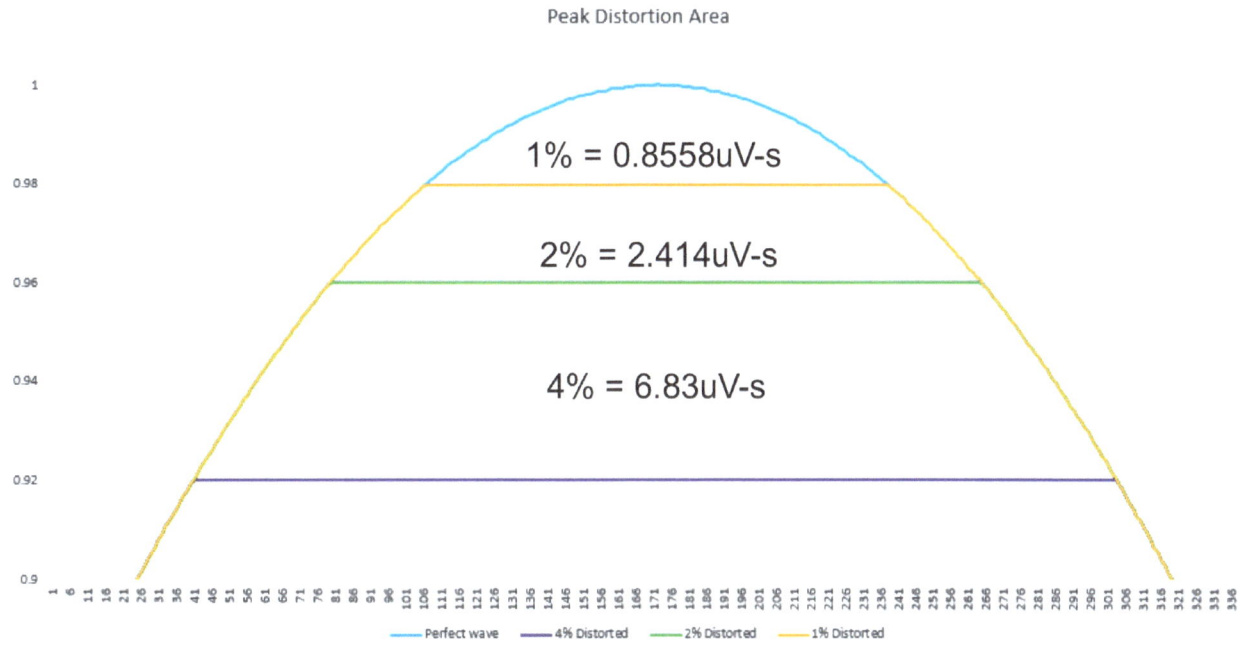

Figure 62 - Peak distortion comparison in area

In Figure 62 you can see that the area removed from the sinusoid peak does not scale linearly because the peak shrinks dramatically as you get closer to the top (or bottom). After integrating the samples that the distortion removed from the sinusoid, the 4% distortion with 1.6uV-s of TDE removed 6.83uV-s of energy from the sinusoid. That is a lot of area, because the sinusoid spends a lot of time in the peaks, slowly humping over and reversing the slope. Now, if it scaled linearly you would expect that the 2%, 0.4uV-s TDE distortion would remove half of that, or 3.41uV-s (based on the 2% vs 4%), or one quarter, 0.4uV-s TDE vs 1.6uV-s TDE, but that's not what happens because of the naturally decreasing area of the sinusoid peak. So instead of the area decreasing by half or a quarter it decreases by a factor of 0.35. It's a rather surprising result. Notice that the same thing happens when we decrease the distortion to 1%, or 0.1uV-s TDE. The area removed from the sinusoid decreases by over half, but not quite one quarter, to 0.3545

Percent Distortion	TDE	Sinusoid Energy	THD	Delta Area/ Delta THD
4%	1.6uV-s	6.83uV-s	-32.1dB	—
2%	0.4uV-s	2.41uV-s	-39.16dB	0.3534/ 7.0dB
1%	0.1uV-s	0.8558uV-s	-46.24dB	0.3545/ 7.1dB

Interestingly, if you compute the Area Delta in dB you get about 9dB, which is 2dB higher than we expected if THD is controlled by area. The reason is simple, even though the peak contains a lot of area, it lacks energy, things are moving very slowly, so area does not translate directly to energy when the distortion is near the peak(s).

In chapter 4 you will see that distortions near the zero crossover create less THD than equivalent distortions near the peak; but *changes* at the zero crossover distortion area have more impact on the THD than similar *changes* near the peak. In fact, near the zero crossover, changes in distortion area scale linearly with the THD. Because of this disparity between Peak vs Crossover distortion, it makes little sense to measure linearity with THD. Since distortions impact the area of a transfer function and area translates into energy, it would make more sense to measure the distortion area (which is easy to do with a linearity test), than apply a sine wave and measure the THD—which is, at best, a clumsy proxy for measuring distortion energy, and at worst, an outright **lie**.

It's important to remember that spectral displays are energy readouts by frequency. The problem is that using a sinusoid to stimulate the transfer function to learn something about it doesn't really make a lot of sense because of the non-linear behavior of sine waves applied against a transfer function. Then there is the noise captured during the test which can also corrupt the measurements of the very small distortion products. Using a linearity test instead, noise can be eliminated by taking the samples over multiple passes, averaging out the noise and increasing the accuracy of the results; thus, non-linearities can be read out directly rather than indirectly with a THD test. It's also a whole lot easier to find a linearity problem in a ramp than counting humps in a spectrum, especially because I am the only person who will color code harmonics for you, although I suspect that if this book gains any traction, some instrument makers might just start doing that for you too.

Earlier, we saw that the number of humps decreases as the distortion gets closer to 90 degrees. Of course, we can't clip at exactly 90 degrees, but we can put a tiny distortion at exactly 90 degrees and see what it does. Are you ready?

Figure 63 - 90 degree distortion spectrum

A flat spectrum—no hump, no curvature at all! Notice that it's a mix of **odd** and **even** harmonics—just like an impulse—because that's what it is, an impulse. That's the only way to make a distortion at exactly 90 degrees. But is it because the single impulse is just an impulse and not because it's at 90 degrees? Remember in chapter 2 we saw that a single impulse on a field of DC (-1V) gave us a perfectly flat spectrum. What happens if we put that single sample width distortion at some convenient place near 90 degrees, say, 88 degrees? The sinusoid will hit it twice, no way around that, but what will the spectrum look like?

Figure 64 - 88 degree distortion spectrum

OK, it's not a flat spectrum anymore, but perhaps it has two humps because a distortion anyplace other than 90 degrees will hit the sinusoid in two places.

You know, we never did look at the single impulse version of our last example clipping wave from figures 60 and 61 which impacted the sinusoid at 78.5 degrees. If the two humps in figure 64 are due to the fact that we have two single width impulses instead of one, then we should also see just two humps. Here goes nothing!

Figure 65 - 78.5 degree distortion spectrum

What do you know? Count them up, 6 humps. Exactly like the 1%, 0.1uV-s TDE clipping example we saw in figure 61, except, of course, this spectrum doesn't roll off so quickly because of the sharp edges. So, the number of humps has nothing to do with the amplitude of the distortion, the energy, the area, or the width. It has to do with the *phase angle* at which the sinusoid intersected the distortion in the transfer function. The closer to 90 degrees, the fewer humps. The further from 90 degrees, the more humps.

But don't get too excited; that's not a really good rule either, because zero crossing distortion at 180 degrees is coming up in chapter 4, and you'll be able to count the number of humps on two hands again, because, well, 180 is evenly divisible by 90 degrees isn't it?

Symmetrical clipping

Now let's try symmetrical distortion—trying to match what we saw with 4% asymmetrical distortion. The problem in trying to replicate that experiment is that we can't just alter 200 points on both the top and bottom of the transfer function and think that it's equivalent to altering 400 points at the top. The reason is that the *area*, that is, the product of time and voltage is what counts in harmonic distortion. The area is not the same for two 200 point triangles as it is for one 400 point triangle.

Think about it. To get the same amount of area, 1.6uV-s of TDE out of the transfer function as we had when we chopped 400 samples out of just one side, we are going to have to chop 400 * sqrt(2) = 565 samples out of the transfer function. It's faulty logic to assume that two triangles made up of 200 samples each will have the same area as one, 400 sample triangle. We are trying to keep the area the same between the two examples of symmetrical versus asymmetrical clipping. In order to compare apples to apples we will need to use the same area, since we know that when we measure harmonic distortion, we are trying to measure energy, and energy in waveforms can be measured in area (volts * time) of the distortion.

So let's get a look at our 5.64% symmetrical clipping distortion in the transfer function. Notice that I had to round it down from 565 to an even number of 282 samples on either side of the transfer function to ensure symmetry; otherwise, what would be the point? Any asymmetry would give us **even** harmonics and ruin the experiment.

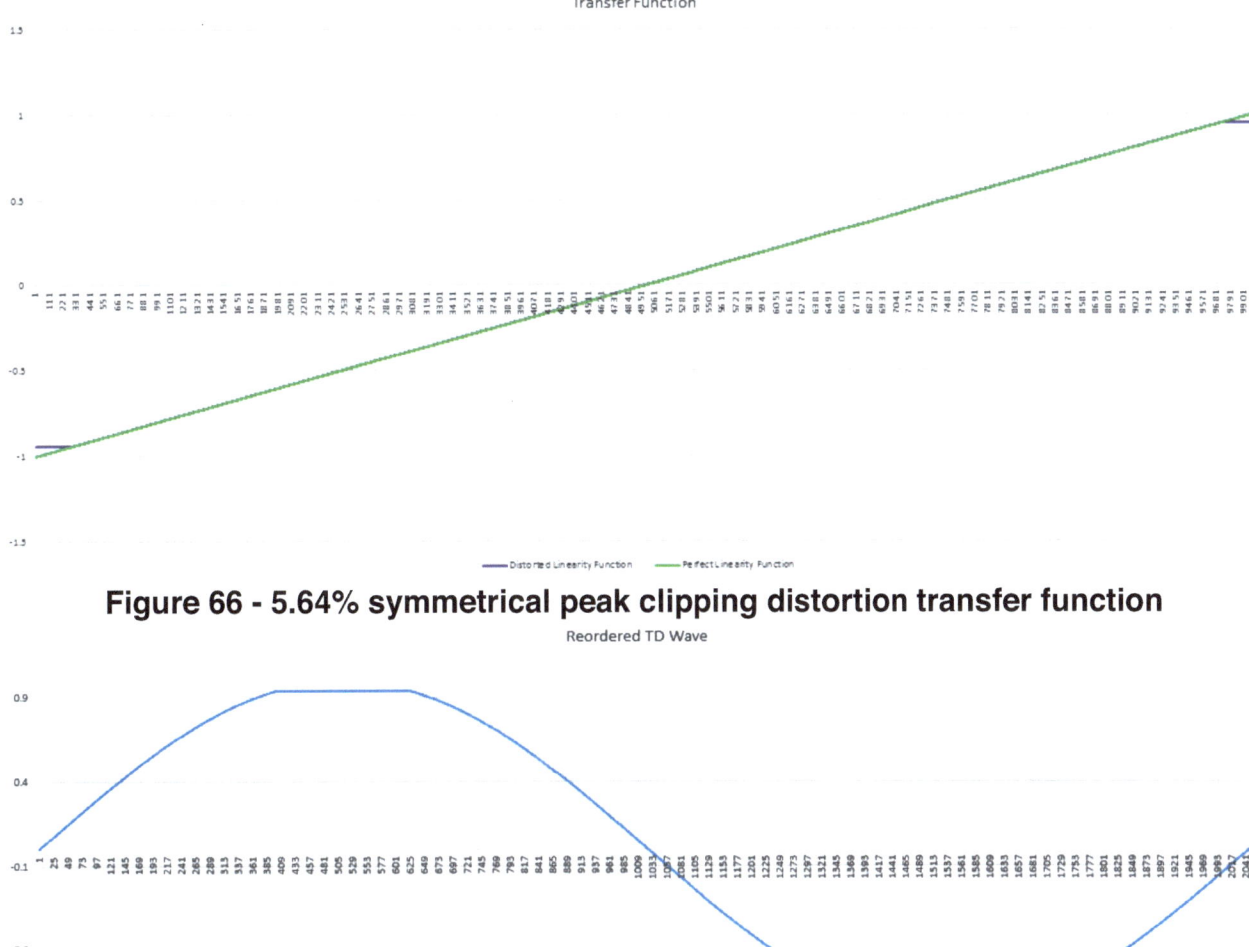

Figure 66 - 5.64% symmetrical peak clipping distortion transfer function

Figure 67 - 5.64%, 1.6uV-s TDE symmetrical peak clipping distortion, time domain

Notice how much of the sine wave is distorted by this seemingly tiny corruption of the transfer function. That's because sine waves spend a lot of time at the top and bottom of the transfer function. That is the beauty of a sine wave: unlike a triangle wave, a sine wave makes a nice smooth transition from a rising slope to a falling slope. A triangle wave does it super fast—in one sample, it's turned around from a rising slope to a falling slope. But a sine wave does a nice, slowly curving turn around, which means it spends a lot of time in those upper and lower reaches of the transfer function which exacerbates the impact of any disturbances at the edges of the transfer function.

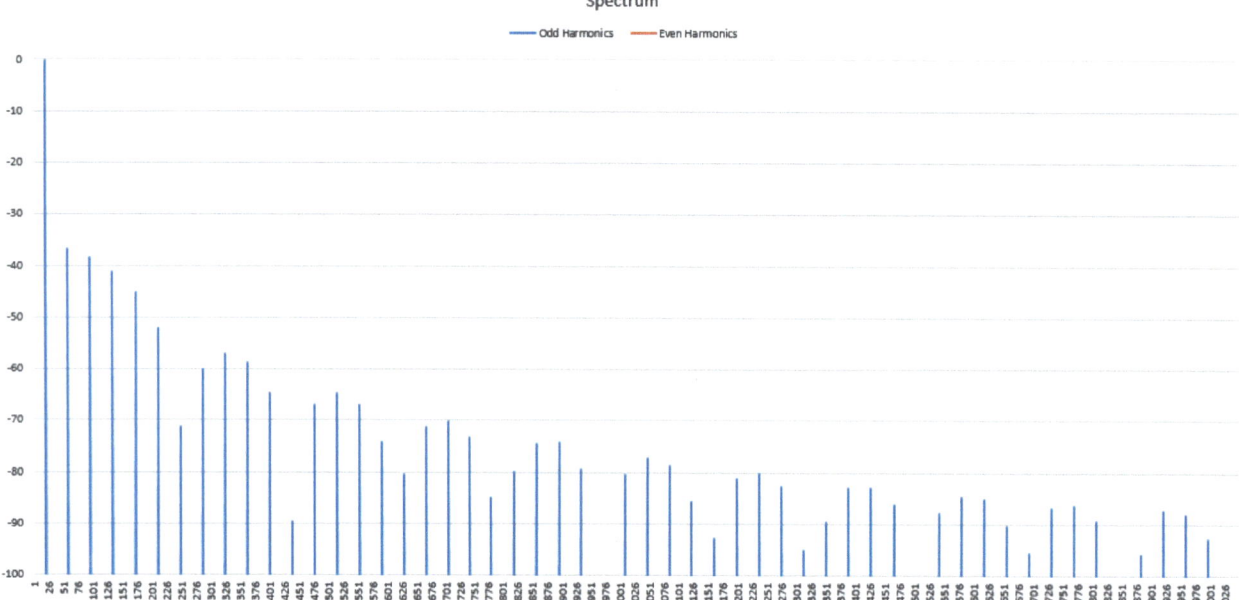

Figure 68 - 5.64%, 1.6uV-s TDE symmetrical peak clipping distortion spectrum

Because we lost the even harmonics, it's a bit harder to see the notches; but, we can see there are minimums. We can count harmonics and find that the first minimum happens at the 13th harmonic. Don't forget to count by twos! These are only odd harmonics, remember? It is rather interesting if you remember that the notch (or minimum) happened on the 11th harmonic when we had 4%, 1.6uV-s TDE clipping on one side only. Not only did the first notch move by 2 harmonics relative to the 4% asymmetrical clipping example, the number of humps is down to 11 from 13 for the 4% asymmetrical clipping example. Since the area is identical for the two examples (1.6uV-s TDE) but the first notch location has changed as well as the number of humps, we must conclude that the number of humps is not a function of area. However, in order to make this 1.6uV-s TDE symmetrical example match the asymmetrical 1.6uV-s TDE example at the beginning of this chapter, I had to set it up so that the distortion intersects the sine wave at 70.66 degrees, passes through 90 degrees, and goes all the way to 109.34 degrees on the top; it also starts clipping at 250.66 degrees, goes through 270 degrees and ends at 289.34 degrees. Remember that we determined that the number of humps decreases as we get closer and closer to 90 degrees, now we've just seen it happen again!

Let's look at what we have seen so far:

Distortion phase	Humps	Symmetry
66.8	13	Asymmetrical
70.66	11	Symmetrical
73.75	9	Asymmetrical
78.5	6	Asymmetrical
88	2	Asymmetrical
90	Flat	Asymmetrical

Notice that it seems to scale nicely—the closer the distortion gets to 90 degrees the fewer humps we get. Notice also that it doesn't matter whether it's symmetrical or not; symmetry makes no difference. Can we deduce a formula from this as we have for the square wave and triangle wave? Absolutely, here it is, just for you:

$$\texttt{VHarmonic = abs(cos(}\pi\texttt{*Harmonic\#*(90-Angle)/180))*TDE}$$

This formula proves that angle is what matters, it explains all the "humpy" patterns you will see when looking at harmonics created by distortion. Note that this formula handles only one simple distortion anomaly, so it can't handle symmetrical distortion anomalies. But now you know that the only difference between asymmetrical and symmetrical distortion is the lack of **even** harmonics. From now on I will refer to this formula as the Bullard Harmonic Solution, or BHS for short, and it can be used by you in an Excel spreadsheet (or Matlab, etc) to reproduce the patterns seen here to match all simple distortion patterns. Again, use $\texttt{(}\pi\texttt{/2-Angle\_in\_rads)/}\pi$ if you are squeamish.

Now back to our example above in figure 68. With 5.64% distortion, spread out as 2.82% on the top and 2.82% on the bottom, the THD goes to -33.2dB for 8 harmonics, and -33.1dB for 50 harmonics. You might think that all those missing **even** harmonics didn't really reduce the THD much, -33.1dB compared to -32.1dB for 4% asymmetrical clipping. With 5.64% symmetrical clipping, we distorted the same area (1.6uV-s) as we did with 4%; and, despite the fact that the **even** harmonics went away, the overall distortion remains about the same: only a 1dB difference between the two. Why is that?

The reason is that when we distort 4% of the transfer function on just one side, we have to go further down the face of the waveform (66.8 degrees vs 70.66 degrees), which takes a much larger chunk out of the top of the sinusoid, versus two smaller chunks of 2.82% taken out of the top and bottom of the sinusoid. So an equal area distortion taken out of the sinusoid in different places causes a differing amount of THD.

Let's take a very slight aside. Remember that to compare symmetrical clipping to asymmetrical clipping we had to equalize the total area to make sure we were comparing apples to apples when looking at THD. But let's say that we temporarily ignore *area* and instead focus on *angle*.

Let's see what happens if we run our experiment with 400 samples of clipping distortion on *both sides*, instead of 400 samples of clipping on just one side. That's double the area, but the angle of the distortion remains the same between the two, since 400 samples clipped off the top and bottom would intersect the sinusoid at the same angle as clipping on just the top of the sinusoid. What would be the result?

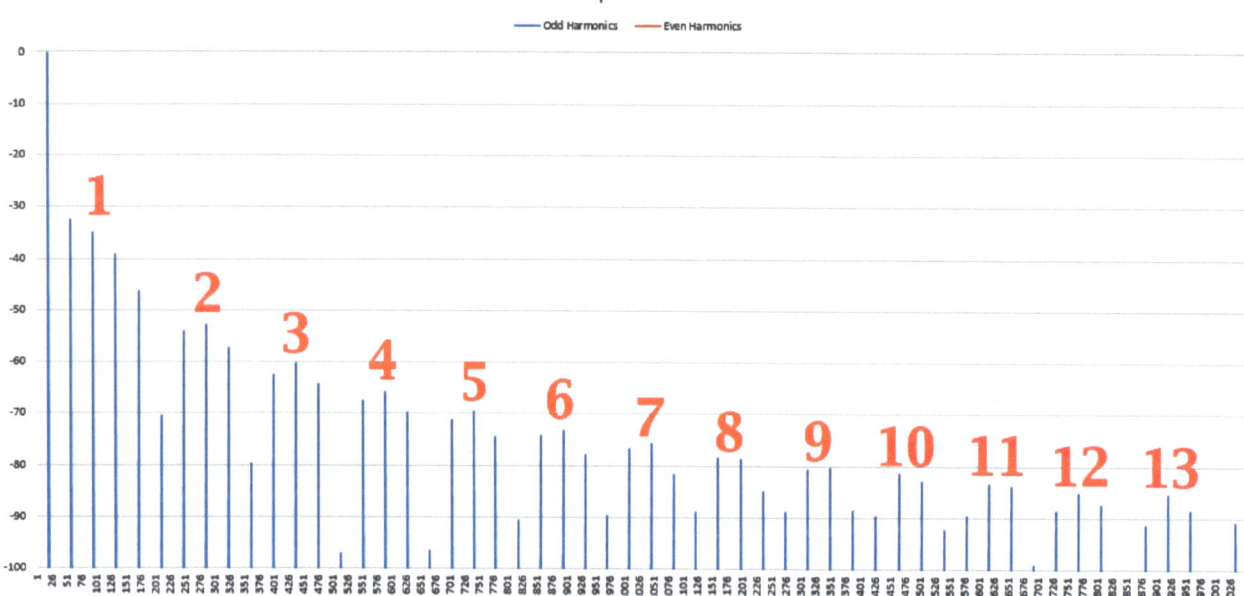

Figure 69 - 4% upper and lower symmetrical peak clipping distortion spectrum

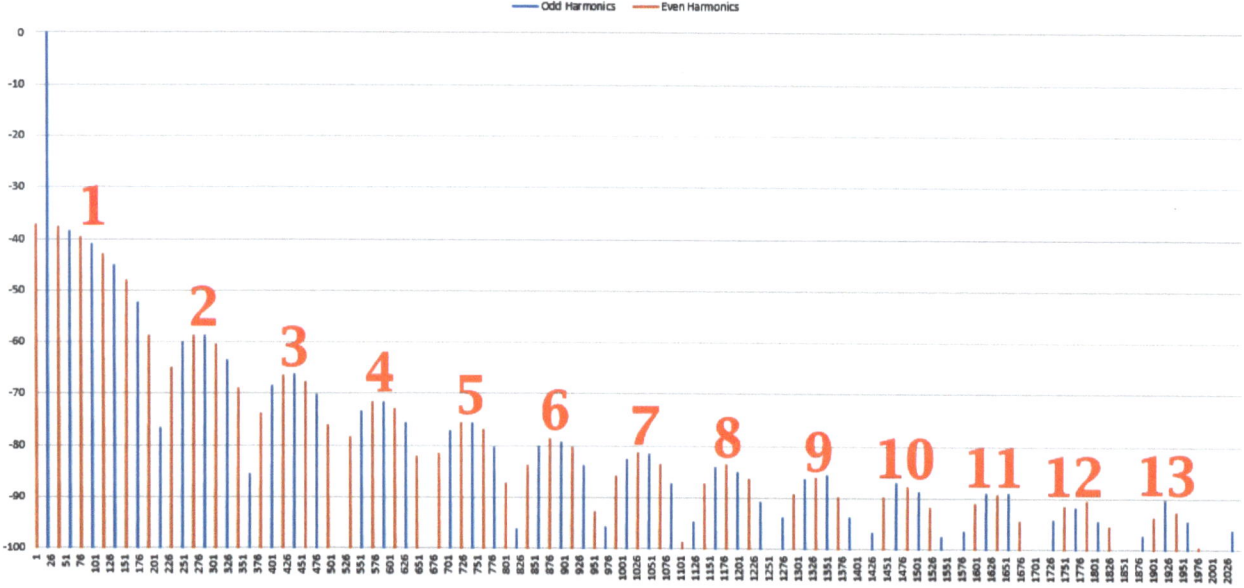

Figure 50 redux - 4% upper peak clipping distortion spectrum

Notice that when we clip 400 samples off the top and bottom we get *exactly* the same number of humps and notches as when we clipped 400 samples off just the top of the sinusoid. The number of humps is controlled by the angle, according to the Bullard Harmonic Solution as we learned earlier. Since both of the distortions in this example happen at exactly 66.8 degrees (and the complementary angle of 246.8 degrees on the lower peak) we are going to get 13 humps—including the extra wide hump near the fundamental. Remember, the angle at which the sinusoid intersects the distortion controls the number of humps in the spectrum according to the BHS. It doesn't matter whether the distortion happens on one side or both sides of the transfer function. However, in the real world it's pretty tough to get the sinusoid to hit exactly the same place on the distortion, which is what I was trying to accomplish in the video I made for Applicos. Varying the DC offset to force the sinusoid to hit the limits of the transfer function accomplished that, but it took some effort. Of course it would have been easier if I had just looked at the mid-level offset value that the linearity test had returned, but then I couldn't have made the movie, or written this book, could I?

You can see that the even harmonics are missing from the symmetrical clipping spectrum in Figure 69, making it a bit difficult to count the actual number of humps. That is why I added the numbers, to make it easier to count them. You can even see that little odd harmonic, the 107th harmonic at the very edge (right side) of the spectrum is in exactly the same place, just a little bit shorter in the asymmetrical spectrum than it is in the symmetrical spectrum by, oh, about **6dB** (-96.3dB vs -90.3dB). You will notice that the amplitudes of the odd harmonics are **6dB** higher in the symmetrical clipping spectrum (fig. 69) than in the asymmetrical clipping spectrum (fig. 50 redux). In fact, they are *exactly* 6dB higher. For example, the 11th harmonic at the first notch is -76.6dB in the asymmetrical example and -70.6dB in the symmetrical example. That should make sense because the total area (TDE) is 3.2uV-s, versus the area of the asymmetrical clipping at 1.6uV-s of TDE. That's twice the area, so the odd harmonics are twice as high. The THD for the symmetrical plot is -29.65dB versus -32.1dB for the asymmetrically clipping wave. But that's only a difference of about 2.45dB. Shouldn't the total distortion energy differ by 6dB, or double what we had because we doubled the area?

No, no, **no**! Yes, the area went up by a factor of two, but all the even harmonics canceled each other out because of the symmetry of the distortion. The odd THD is -29.65dB and if the even harmonics were not canceled out, the even THD would be about the same, yielding a THD of about -26dB versus -32.1dB, a difference of 6dB. Twice the distortion for twice the area. Because the even THD is -120dB the even harmonics add virtually nothing to the sum total of harmonic distortion. Oh yes, they are there in one half cycle but then they are canceled out in the next. There is no way to actually prove they are there unless you remove the distortion on either the negative or positive side—then they suddenly appear. The closest we have gotten to seeing them was to note that the even harmonics matched the pattern of 1/harmonic# in a 100% Tr/Tt triangle wave like we normally get with a 50% duty cycle square wave.

Because the distortion I added actually started at 66.8 degrees into the cycle, due to the fact that my distortion compromised the last (and first in the case of symmetrical distortion) 400 samples of the transfer function, the number of humps and notches in the spectrum are identical and are predicted by the Bullard Harmonic Solution. Once you get good at reading spectra, you can discern the location of the distortion simply by counting the number of humps and notches in the spectrum. I doubt you'll ever see that much distortion in your career, and sometimes it's really hard to get a good count, as in the case of symmetrical distortion when the missing even harmonics make it very difficult to count humps and notches. Perhaps some instrument company will one day write some code to calculate the phase angle of the distortion and thus its location in the transfer function by using the BHS in reverse. Still, this explains why the number of humps and notches change as I change the amount of distortion. The less distortion I add, the fewer samples that get corrupted, the higher along the sinusoid the distortion happens, and so the fewer humps and notches. The amplitude of the harmonics depends on the total area disturbed, but the number of humps and notches depends on where in the transfer function the distortion intersects the sinusoid.

Now, back to our experiments with symmetrical clipping. As before, we lower the distortion down to 2.82%, which is the symmetrical equivalent (in area) of 2% asymmetrical clipping.

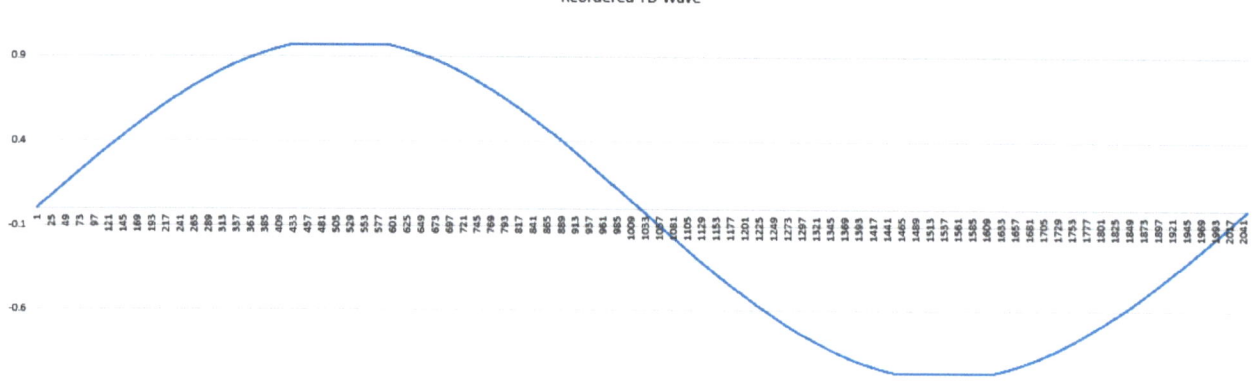

Figure 70 - 2.82%, 0.4uV-s TDE symmetrical peak clipping distortion, time domain

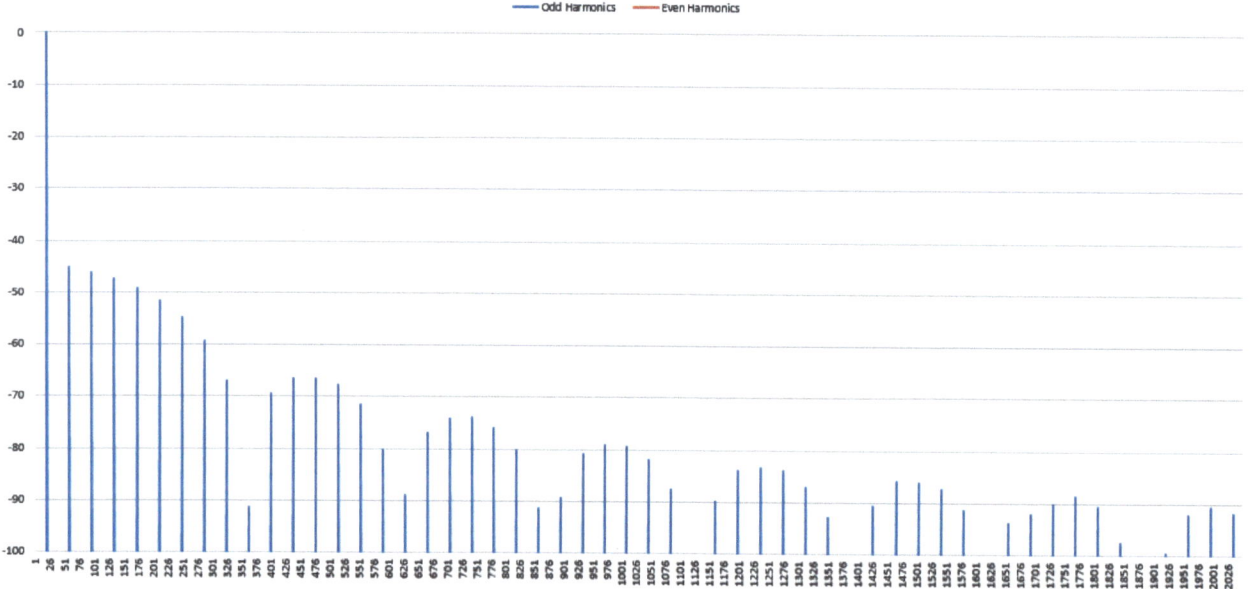

Figure 71 - 2.82%, 0.4uV-s TDE symmetrical peak clipping distortion spectrum

The THD in this case is -40.68dB for the first 8 harmonics and -40.08dB for the first 50 harmonics. The **odd** THD is -40.08dB, and the **even** THD is -120.15dB. The later is virtually nonexistent, which explains why the **odd** THD equals the 50 harmonic overall THD number. Because the area is one quarter of what it was in the 5.64% example, you would expect the THD to be down by 12dB; but, it's only down by about 7dB from the 33.1dB value for the 5.64%, 1.6uV-s TDE waveform. While overall distortion near the peaks creates more THD, the changes in THD are not linear with increases or reductions in distortion area as we saw before. Changes in distortion at the peak that should result in a 12dB change in the THD result only in a 7dB change in THD. Not an intuitive result, but there it is!

Moving along, we reduce the distortion again and look at the result:

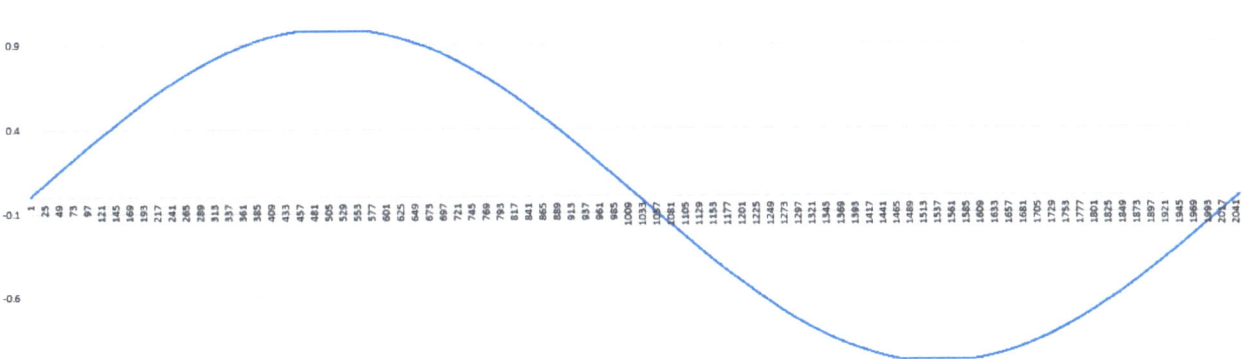

Figure 72 - 1.4%, 0.1uV-s TDE symmetrical peak clipping distortion, time domain

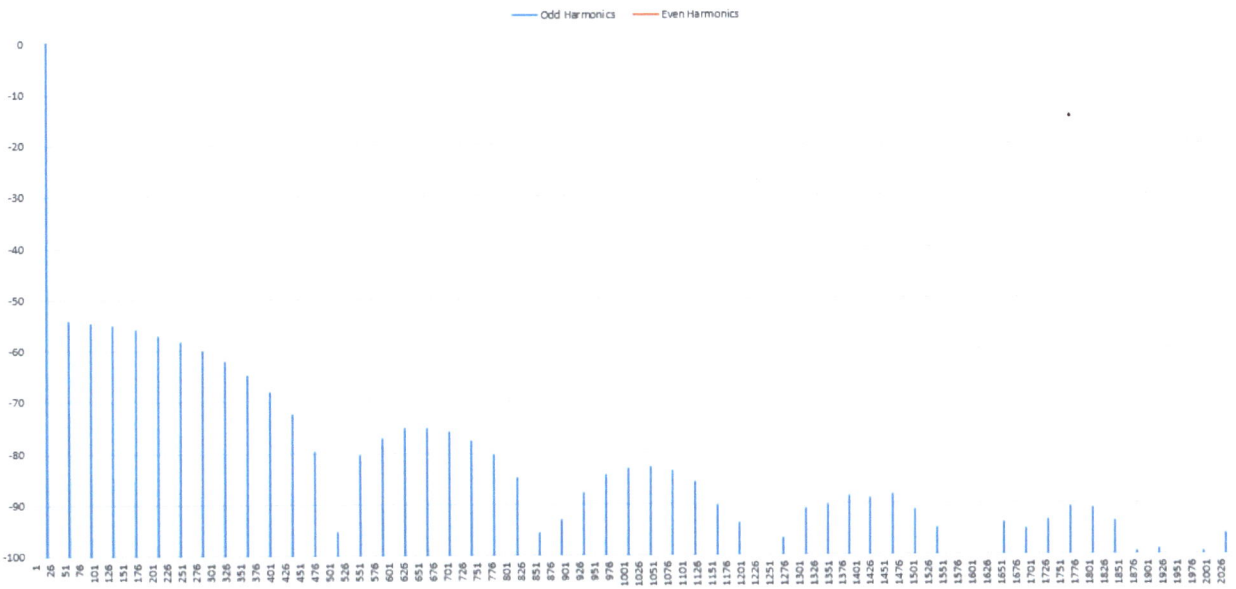

Figure 73 - 1.4%, 0.1uV-s TDE symmetrical peak clipping distortion spectrum

Now the THD has dropped to -48.87dB for the first 8 harmonics and -47.28dB for the first 50 harmonics. Virtually all the harmonic energy is in the odd THD at -47.28dB. Even THD is still at -120.2dB, which is very, very low.

How to recognize Peak distortion

Notice that in every example we've seen in this chapter, the harmonics, both odd and even (when there were even harmonics) started out at a high level and then followed a decaying trajectory, humping over and over in a sine(x)/x fashion. When you see this spectrum, you might think there is a low pass filter at work; but this is not the case. The reason for the roll-off is that the top and bottom of the wave being lopped off is a low frequency distortion with no sharp edges. The cosine function in the BHS is responsible for the behavior of humps and notches; and, as we saw, it was the actual *angle* where the distortion impacts the sinusoid that defines how many humps and notches we get according to the Bullard Harmonic Solution, just like the sine function and the duty cycle (or H2L transition angle) was responsible for the humps and notches in a square wave spectrum.

Again, realize that your particular spectral data from whatever instrument you use will determine how many humps appear, based on the sampling frequency it uses. The pattern will also be influenced by any anti-aliasing filters you might have applied to the signal before it gets digitized, because so many engineers fear aliasing, which is really your friend[19], not your enemy despite what MIT and other ignorant professors and engineers might tell you. But one thing will not change: that is the fact that with peak clipping, the harmonics always start out high and decay down to a notch, either fast or slow, depending on how much of the wave is clipped—which controls the angle of sine wave where the clipping occurs, **not** how much area is distorted. This roll off behavior is not predicted by the Bullard Harmonic Solution, because all it can do is give you the pattern of humps and notches; it cannot account for the shape of the distortion, only the behavior of the spectrum for a single distortion at a single location in the transfer function. However, I suspect that if I was to integrate the area of the transfer function over one degree, i.e. integrate all the energy in the linearity curve in 1/180th of the transfer function, starting at the bottom (for 270 degrees) and apply that number to the Bullard Harmonic Solution, then repeat 179 times and integrate the results, I would likely accurately replicate the actual spectrum. It could be a fun experiment for someone who wants to solve this issue.

As we will see in the next chapter with Crossover Distortion, the *spectral signatures* do not look like the spectra in this chapter at all. The pattern changes depending on where the stimulating sinusoid hits the distortion. So, not only do the number of humps and notches tell us at what angle the distortion intercepted the sinusoid; also, the pattern of harmonics changes as well, and, it is very different from what we have seen so far.

---

[19] http://www.danbullard.com/dan/test_fast_clocks_page_1.html

Chapter 4
Distortion Signatures - Crossover

After Chapter 3 you might think you have a pretty good idea about how harmonics are created by distortion. You would be partially right; but don't skip this chapter, because there is a big surprise lurking here that could lead you down a seriously wrong path when you come across a Class B or Class AB amplifier.

In audio and linear RF amplifiers, nothing beats a Class B, or its "perfectly linear" variant, the Class AB amplifier. This topology allows a push-pull transistor pair to source or sink AC current directly to a speaker or antenna (in the case of RF amplifiers).

Class AB is a significant improvement over Class B, because in true Class B amplifiers, there is a small portion of the transfer function where both transistors are off, which reduces the output current and voltage to zero. Class AB fixes this by applying a small bias voltage to keep at least one transistor on at all times.

But Class AB suffers from the fact that even if one transistor is on at all times, a transistor that is barely conducting is still not in the most linear portion of its transfer curve. This can cause strange distortion maladies that confound the engineer's intent to secure the best quality output for his application.

Before we get started, let's take a look at what happens when our transfer function distortion moves from the peak towards the zero crossover. In the following table I show how the 50 harmonic THD decreases as a 1% or 100 sample, 0.1uV-s TDE distortion travels down my transfer function. At the very top, the 0.1uV-s distortion intersects the sinusoid peak and covers everything from 78.5 degrees to 101.5 degrees, just like it did when we ran this experiment in figures 60 and 61.

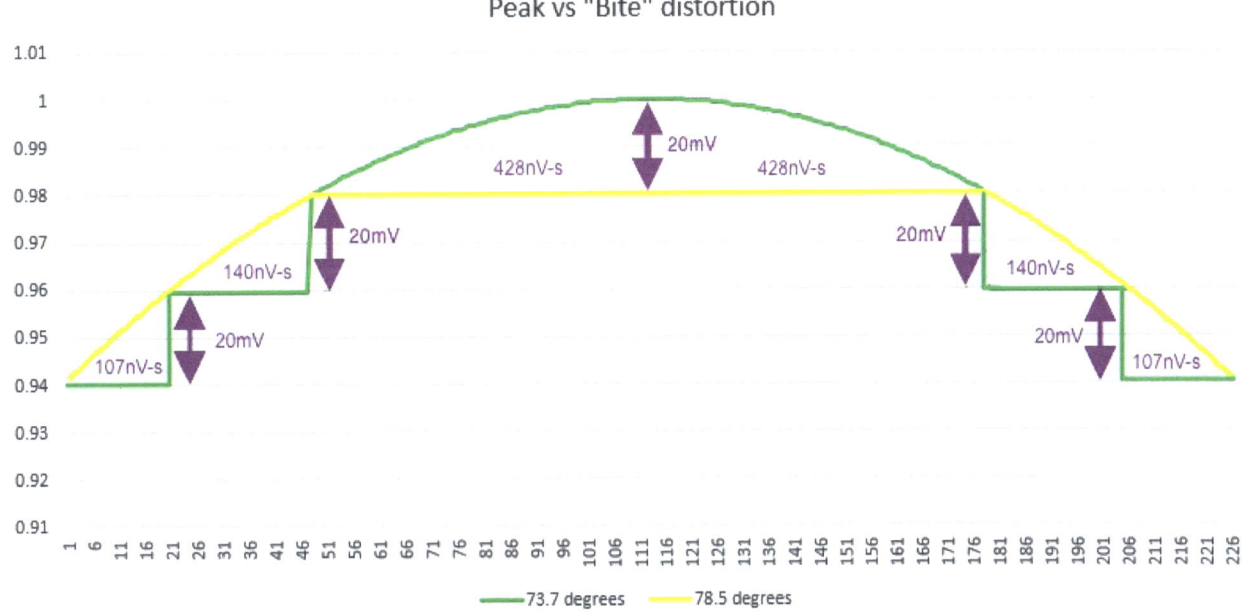

Figure 74 - Peak vs Bite distortion along a sinusoid

At the peak, the distortion impacts 131 samples of the applied sine wave, which in essence forms two back-to-back triangles 65 samples wide, with rounded hypotenuses of course, exactly 20mV high at the peak. By integrating the missing area we find that 855.8nV-s is missing from the sine wave peak, just like we saw in figure 62. In an undistorted wave, a total of 13 points is spent at the actual peak value of +1.0, but lucky for us, because it's not moving it's not creating any energy. In fact, we know from our *harmonic signature* that the harmonic content of the peak is concentrated in the lower harmonics.

Angle	Area	THD	Area Delta/ THD Delta
78.5	855.8nV-s	-46.24dB	NA
73.7	280nV-s	-52.72dB	0.33 / 0.47
70.1	214nV-s	-54.42dB	0.77 / 0.82
66.9	179nV-s	-55.7dB	0.84 / 0.87
64.2	153nV-s	-56.95dB	0.86 / 0.87
61.6	134nV-s	-58.04dB	0.87 / 0.88
59.3	130nV-s	-58.2dB	0.97 / 0.98
57.1	124nV-s	-58.52dB	0.96 / 0.96
55.1	125nV-s	-58.48dB	1.0 / 1.0
53.1	115nV-s	-59.06dB	0.92 / 0.94

Move the distortion down 100 sample points in the transfer function to where it first intersects the sine wave at 73.7 degrees and the sinusoid only hits 27 points out of 100, but of course it hits them twice, just like the peak did (frontside+backside). Each triangle represents 140nV-s of energy. This is *real* energy, stolen from the sinusoid, not like TDE which is stolen from the transfer function, kind of like the difference between potential energy vs real energy. TDE produces no real energy because it takes an activation signal to tease the energy out of the transfer function, just like the marble in your cup which creates no motion without orbiting the circumference of the cup. In the case of a THD test, we use a sine wave, in a linearity test, we use a ramp wave. The ramp wave has the advantage of being linear, and it hits every point in the transfer function, unlike a sine wave.

Since there are two, 140nV-s triangles, the total energy for our second entry in the table above is 280nV-s, far smaller than the 855.8nV-s of energy stolen at the peak, but because the peak contains very little energy, and most of that is concentrated in the lower frequencies, the *harmonic signature* for these "bite" distortions are quite a bit different as you will see later in this chapter. This large difference in energy between "peak" and "bite" distortions explains why there is a large difference in THD between the first entry of this table versus all the others, including all forms of crossover distortion.

As we move the distortion down the transfer function, fewer points are "hit," or impacted, shortening the base of the triangle making each one smaller and smaller, reducing the area and therefore the energy. When the energy goes down, so does THD in a semi-linear fashion. Notice though, as we get closer to the sinusoid's almost linear rise (or fall) near 45 degrees (or 135 degrees), we see less and less impact, but a closer match (more linearity) in the area delta to the THD delta. Since the face of the sine wave from 315 degrees to 45 degrees, and from 135 degrees to 225 degrees is almost linear, there is no point in extending this table. A little later you will see a repeat of this exercise with a zero crossing distortion of 0.1uV-s TDE exhibits a THD of -64dB. That means that from our last entry in the table above to 0 degrees we improve THD by only 5dB, nothing compared to the staggering difference between the top entry in this table and the last entry of 12.88dB. It's worth noting the case of two entries, the one at 57.1 degrees and 55.1 degrees. Notice that both impact the same amount of energy and so THD numbers are almost identical, further validating my claim that THD is caused by area.

Asymmetrical Crossover Distortion

Let's start by looking at some asymmetrical crossover distortion failures and see how they impact the harmonic content of the output signal. Don't expect a repeat of the clipping distortion that we saw earlier; crossover distortion has a different signature which we touched on back in chapter 1, although I can assure you that the rule we found for **odd** and **even** harmonics still prevails. In fact, because of the sharp edges, the Bullard Harmonic Solution almost perfectly predicts the shape of the spectra to come.

As before, we will start with 4% Asymmetrical distortion. First, in the Time Domain, you will notice a tiny notch taken out of the sinusoid just above the zero crossing at 4.6 degrees:

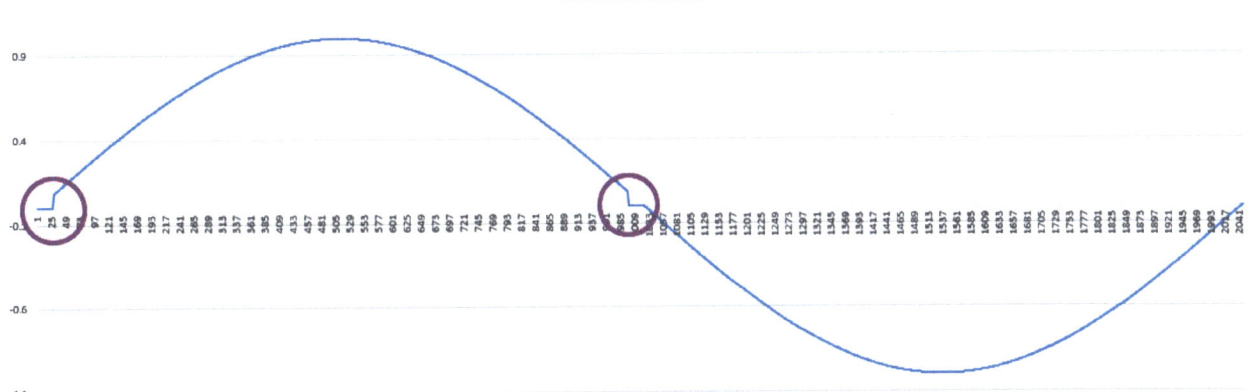

Figure 75 - 4%, 1.6uV-s TDE asymmetrical zero crossing distortion time domain

Because the area impacted is so small compared to the same level of distortion at the peak, you can expect that the THD is much lower. I made sure to circle the distortion in violet so you can clearly make it out.

The impact of non-linearities near the center of the transfer function has less effect because the area affected is smaller than a similarly sized non-linearity at the edges of the transfer function. The one time this is not true is when a design refrains from using the transfer function near the edges, to avoid outright clipping. Non-linearities that are never hit, like speed bumps on a road nobody ever drives on, have no impact on the signal.

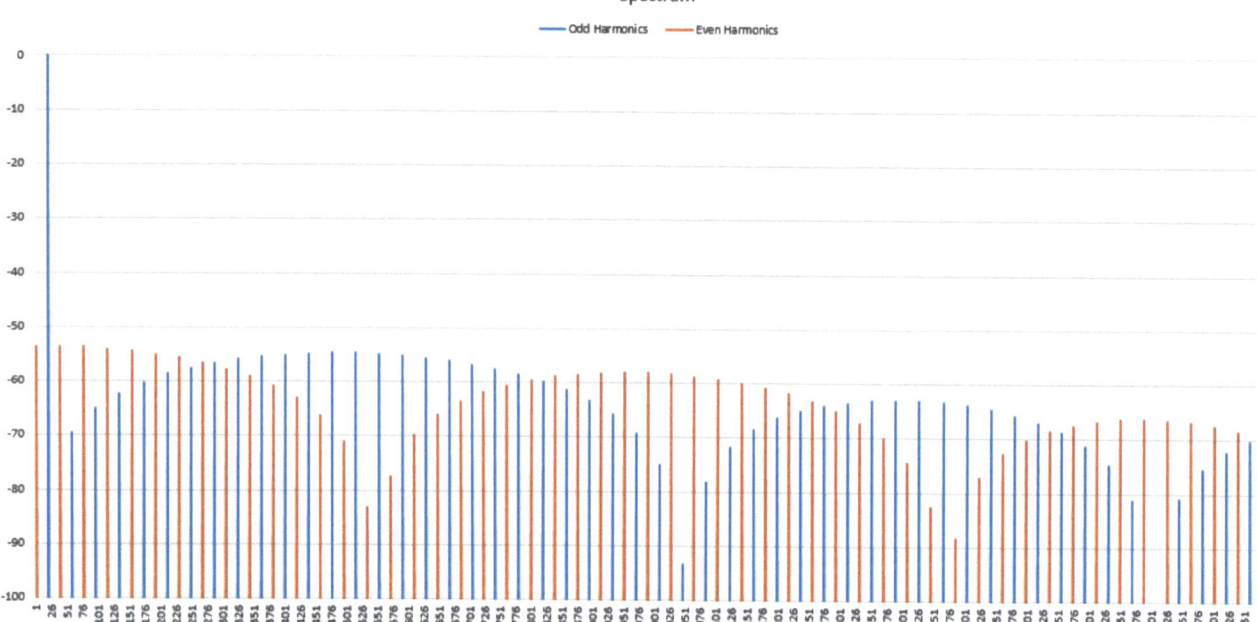

Figure 76 - 4%, 1.6uV-s TDE asymmetrical zero crossing distortion spectrum

The THD is -47.45dB for 8 harmonics and -40.83dB for 50 harmonics. Quite a difference, and you might wonder why. Did you notice that the **even** harmonics start out fairly high at the lower frequencies and then roll off as they did when we clipped the peak of the sinusoid? But notice also that the **odd** harmonics start off quite low and then rise up out of the noise floor to peak quite a way out, around the 27th harmonic at about -54dB. If this waveform were an A4 tone at 440Hz, the 27th harmonic would be 11,880Hz, well inside the range of normal hearing. If you only look out to the 9th harmonic, the **odd** harmonics never even get above -60dB. This might lead someone to believe that the harmonic distortion of this waveform isn't all that bad.

Now you know why I look all the way out to the 50th harmonic, and even that's not enough! Once you get away from the peaks, the harmonics shoot off to the right and your stationary "goal post," 5 harmonics, 7, 9 or even 20 harmonics, will not be high enough to catch all of them. This behavior is a classic crossover distortion signature; but, unfortunately, most people don't realize that **odd** harmonics behave this way in crossover distortion.

Let us revisit the statements made by Bob Metzler in his classic book, The Audio Measurement Handbook. On page 29, he said "non-linearities which are not symmetrical around zero (...) produce dominantly even harmonics." Why did he say this?

Add a filter that removes every harmonic above the 9th (4.4KHz with a 440Hz fundamental), and the reason becomes clear:

Figure 77 - 4% asymmetrical zero crossing distortion spectrum filtered

Once you remove everything above the 9th harmonic it appears that asymmetrical crossover distortion creates mostly **even** harmonics. Little did the Audio Precision people realize that had they not ignored harmonics above the 9th, they would have seen that **odd** harmonics are generated with amplitudes equal to the **even** harmonics, just a little further out in the spectrum. In other words, the Audio Precision folks didn't look far enough out. Even if you make the fundamental 1KHz. it still makes sense to look past the 9th harmonic, since human hearing extends up to at least the 20th harmonic. But again, most engineers *assumed* that harmonics roll off the way we saw back in chapter 3 when we were clipping the peak(s) of the sinusoid.

This is what makes harmonics so fascinating to me. The first *notch* or null in the **odd** harmonic spectrum happens near the fundamental when the non-linearity occurs at the zero crossing. We saw in chapter 3 how the patterns of humps and notches were influenced by the *phase angle* of the sinusoid that stimulated the transfer function according to the BHS. Now we see that while the **even** harmonics start out high near the fundamental, the **odd** harmonics' first notch appears right at the fundamental and then rises to form a peak well beyond the 9th harmonic in many cases. The assumption that all harmonics roll off is bogus, and I (and now **you**) are among the few who understand this. Look at any amplifier spec: virtually no one looks beyond the 9th harmonic for THD. But, as you can see here, that is crazy! The **odd** harmonics don't roll off when the distortion happens near the zero crossing; they ramp up! Ignoring harmonics above any arbitrary number is insane! Wait until we get to symmetrical crossover distortion! You'll come the conclusion that there are a lot of amplifiers out there that pass the THD spec, but still sound really bad!

So, does the Bullard Harmonic Solution correctly predict an asymmetrical crossover distortion at 4.6 degrees? You decide:

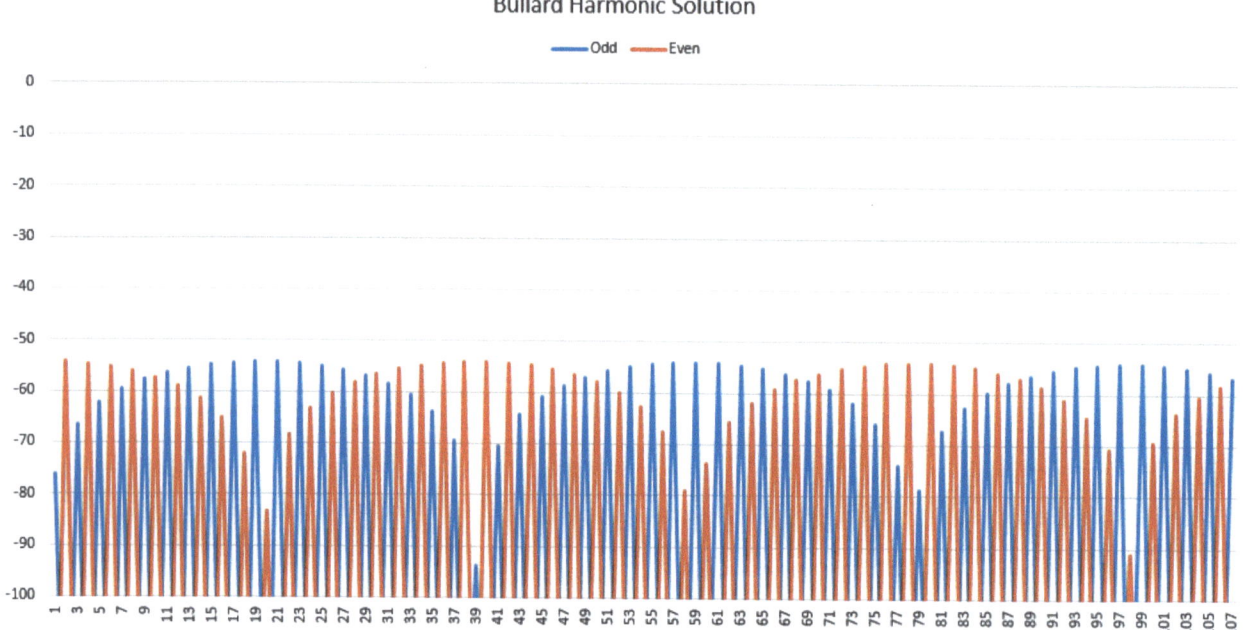

Figure 78 - 4.6 degree asymmetrical distortion BHS spectrum

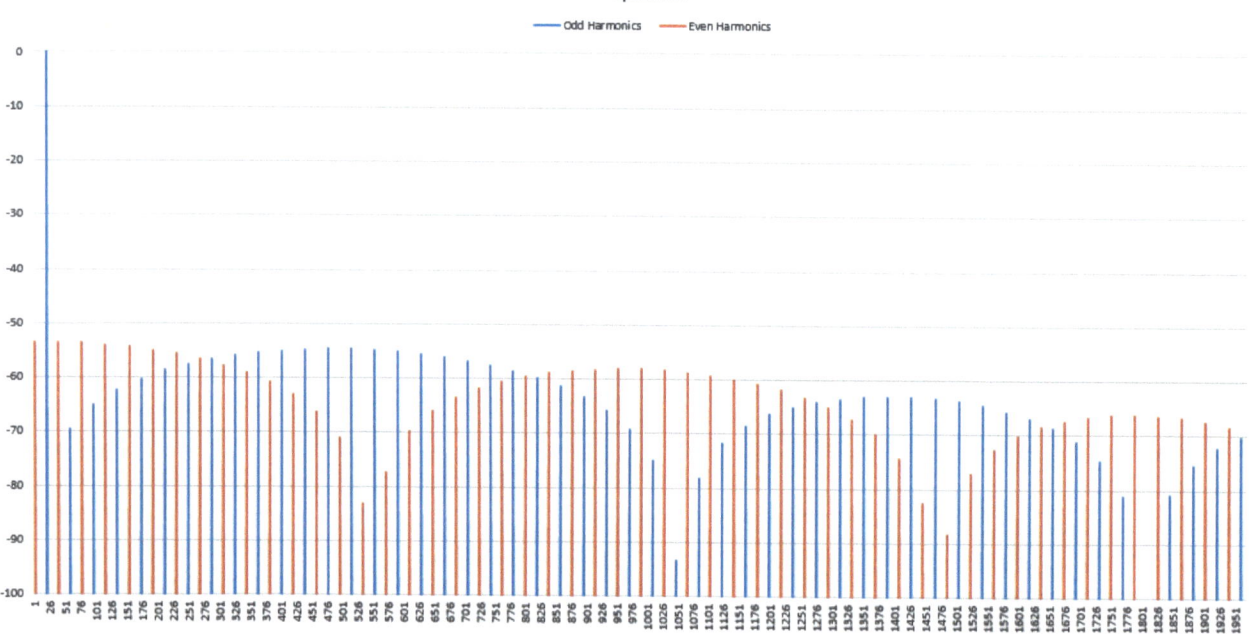

Figure 76 redux - 4%, 1.6uV-s TDE asymmetrical zero crossing distortion spectrum

At first glance it seems that the Bullard Harmonic Solution matches the actual spectrum pretty well. However, on closer examination, it doesn't match all that well; the first notch occurs at the 28th harmonic (14th **even** harmonic) in the real spectrum, but the BHS predicts that it should have happened at the 20th harmonic (the 10th **even** harmonic). What's wrong?

If we revise the distortion to set just a few points to -1V at exactly 4.6 degrees, instead of trying to simulate crossover distortion, we get a better match.

Figure 79 - 4.6 degree asymmetrical zero crossing distortion spectrum

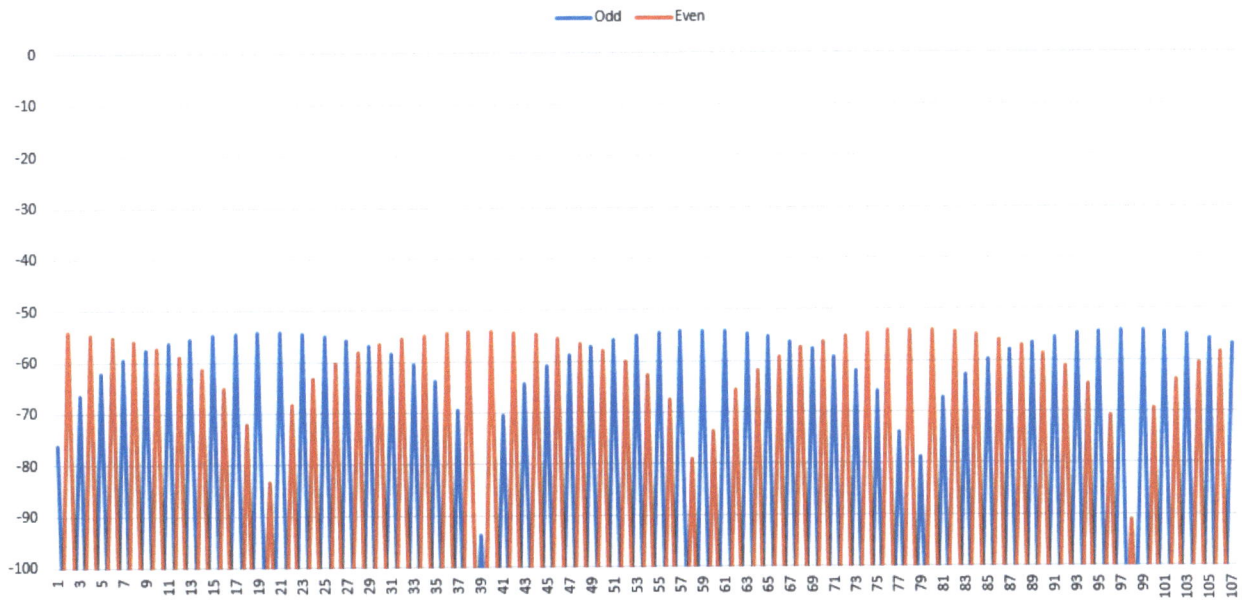

Figure 78 redux - 4.6 degree asymmetrical distortion BHS spectrum

Notice that this one matches much better—in fact almost perfectly! The reason that the 4.6 degree crossover distortion didn't match as well is that 4.6 degrees is where the distortion *ended*. Because my simulated asymmetrical crossover distortion forms a triangular shaped *bite* taken out of the sinusoid, the average phase angle of the distortion is closer to about 3.7 degrees rather than the ending location of 4.6 degrees.

So, let me program that into my formula and see if that more closely matches my simulated crossover distortion:

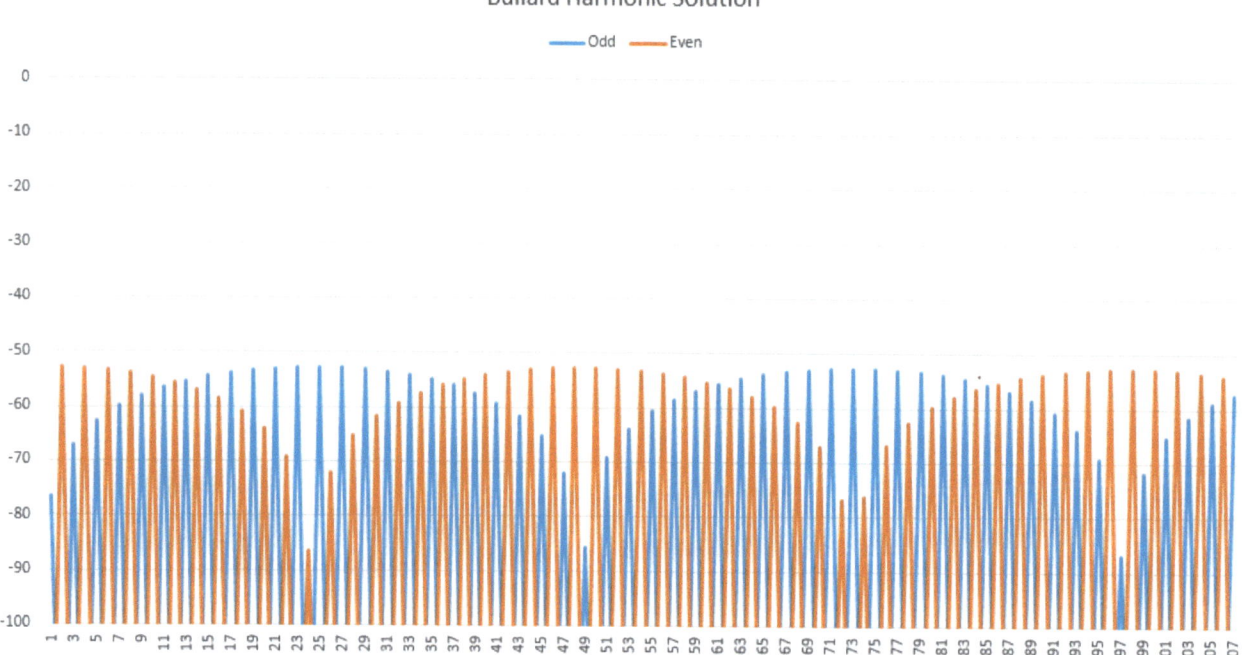

Figure 80 - 3.7 degree asymmetrical distortion BHS spectrum

Now let's compare it again with the spectrum of the simulated asymmetrical crossover distortion at 4.6 degrees:

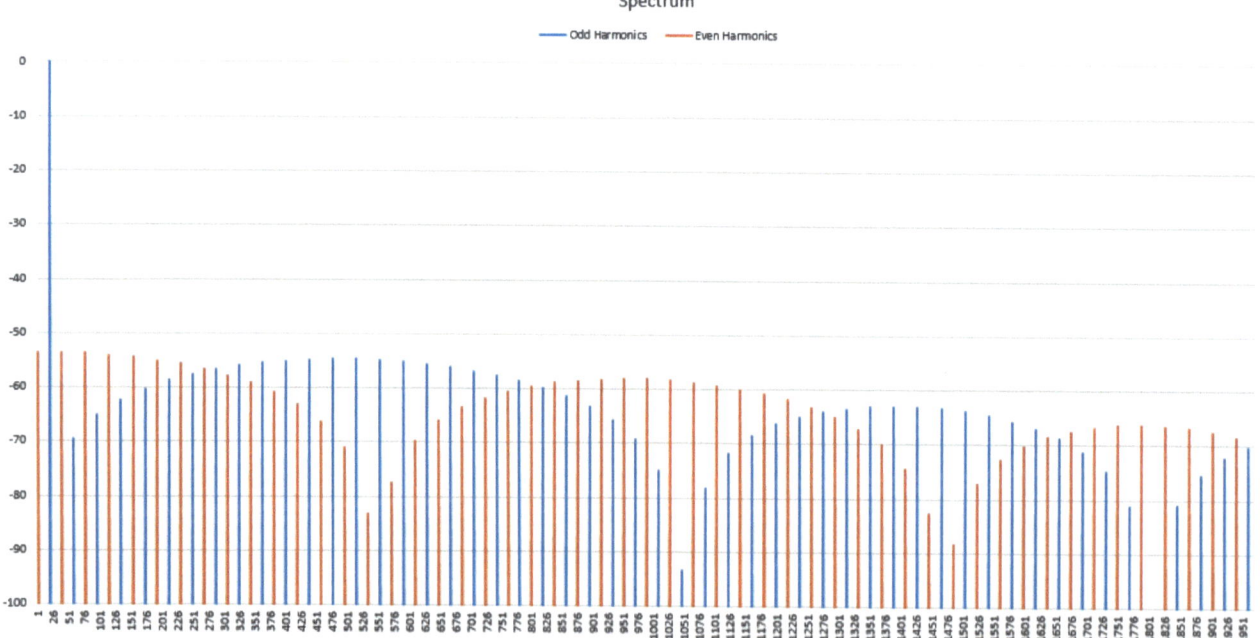

Figure 76 redux - 4%, 1.6uV-s TDE asymmetrical zero crossing distortion spectrum

This looks like a much better fit. Notice that the first notch is closer, but still in the wrong place. It's off by 4 harmonics (24 vs 28 for the real thing); but, otherwise, the shape of the humps, the number of humps, etc. seem to match pretty well. This is to be expected, because the Bullard Harmonic Solution only handles one phase angle; and, near the zero crossover we are going to be forced to rely on the average phase angle. Remember that the BHS worked remarkably well near the peaks, except of course for the roll-off due to the fast drop off of harmonics from the slow rate. The BHS pretty much assumes that we are dealing with a short duration glitch in the transfer function. You are welcome to try to improve it as I suggested in the last chapter, and I would love to see derivations that cover it all. But, for now the Bullard Harmonic Solution is the best we are going to get, and it is far better than any other proposal offered so far.

One more thing: notice that the fundamental is missing because the BHS only predicts distortion products, not the fundamental, unlike the formulas I gave you for square and triangle waves.

Moving along, let's reduce the severity of the nonlinearity to 2% (200 samples forced to zero out of a 10,000 point transfer function) with a TDE of 0.4uV-s and see what the effect is:

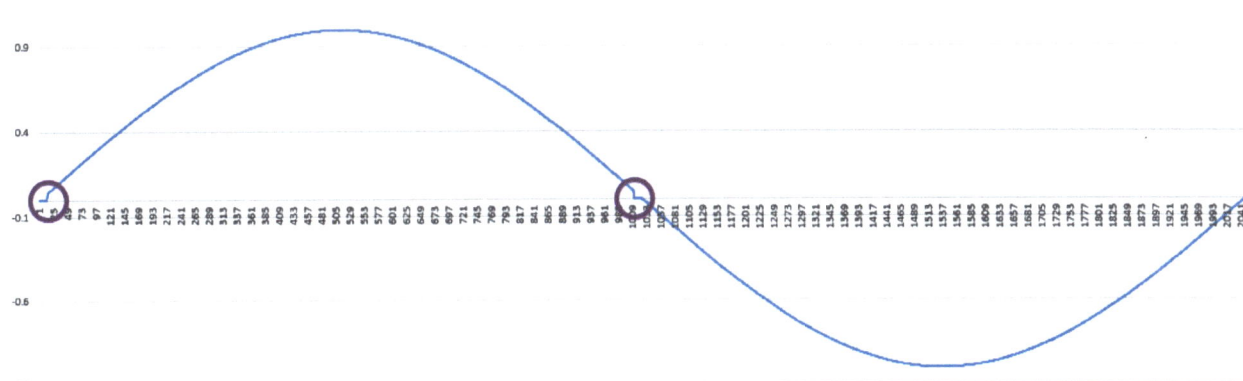

Figure 81 - 2%, 0.4uV-s TDE asymmetrical zero crossing distortion time domain

As before, I highlighted the distortion in violet because it is quite difficult to see. It appears twice on the time domain plot, of course, because the sine wave has to travel though the distortion twice. That does mean that the harmonic energy is double, just as when the sinusoid went through both the rising edge of the positive peak and negative edge of the positive peak forming a back-to-back triangle in the positive peak clipping example in figure 74. The sinusoid must pass through all non-linearity twice, in all cases. There is no amplifier topology I know of that uses four transistors, one for each polarity and one for each slope! Any non-linearity will be hit on the way up, and the way down, just as in the old saying, what goes up, must come down.

Now for the spectrum:

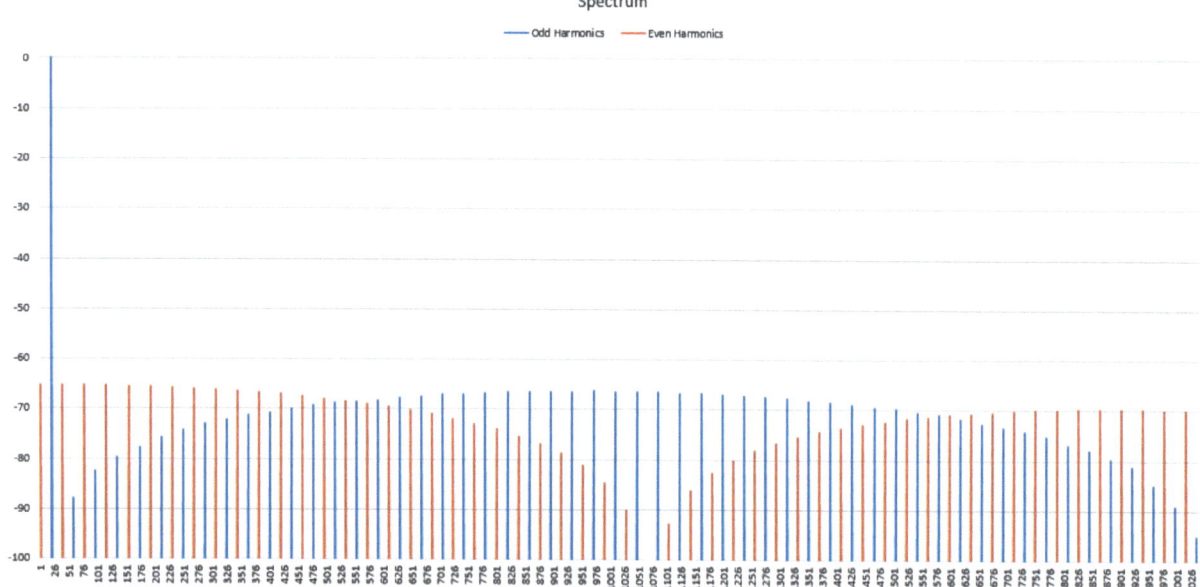

Figure 82 - 2%, 0.4uV-s TDE asymmetrical zero crossing distortion spectrum

The THD is -59.22dB for 8 harmonics and -51.56dB for 50 harmonics. Again, there is a large disparity between the THD for 8 harmonics versus 50 harmonics. If we look at the 50 harmonic **odd** THD, it's -55.1dB, and the **even** THD is -54.08dB.

I can almost forgive Audio Precision for thinking that asymmetrical distortion creates mostly **even** harmonics—but then I couldn't smile with the sense of superiority that you can now share.

Very quickly, let's try looking at the output of the BHS using a phase angle of 1.68 degrees instead of the end of the distortion at 2.29 degrees. This would approximate the average phase angle of the distortion.

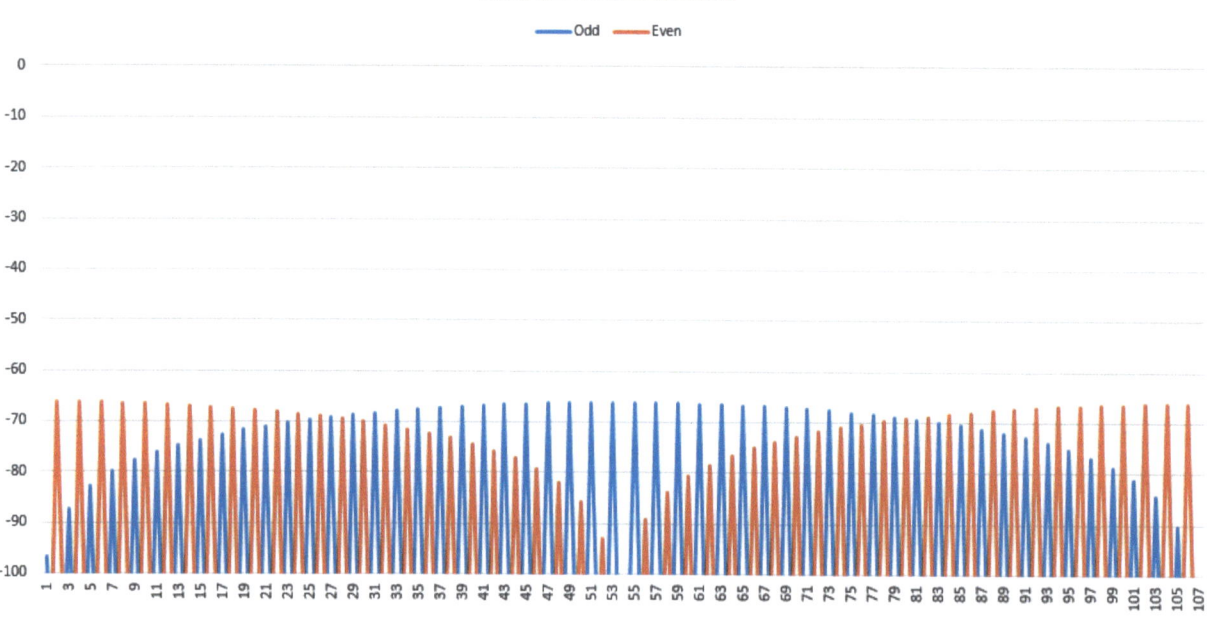

Figure 83 - 1.68 degree asymmetrical distortion BHS spectrum

A pretty good match, no? Because the average of the total area affected by the distortion does not occur at the beginning of the distortion (1.68 degrees vs 2.29 degrees), the Bullard Harmonic Solution can only get you close to the problem. But, then again, before you started reading this book, you had no idea that area or phase angle had anything to do with harmonics created by distortion. The BHS is a whole heck of a lot better than the spurious (and false) generalities that came out of Audio Precision or anyone else, including your professors.

Moving on, once again let's lower the amount of distortion and see how it impacts the spectrum—this time to 1% asymmetrical distortion with a TDE of 0.1uV-s:

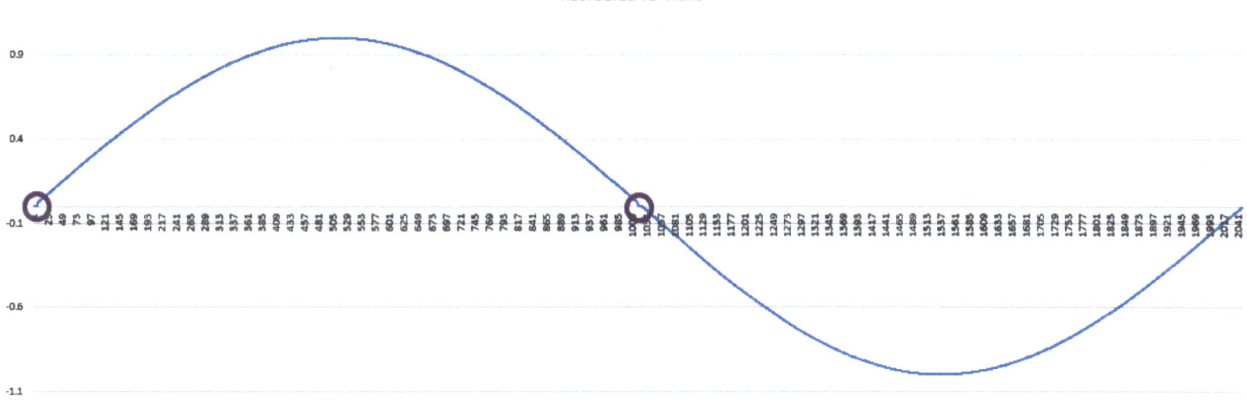

Figure 84 - 1%, 0.1uV-s TDE asymmetrical zero crossing distortion time domain

This distortion is very, very small, almost impossible to see if you didn't know it was there. This is why we no longer employ humans to sit and stare at oscilloscopes all day long to verify the quality of audio amplifiers. We found that doing spectral analysis gave us more information with very little cost. If only we had interpreted the results more accurately. C'est la vie...

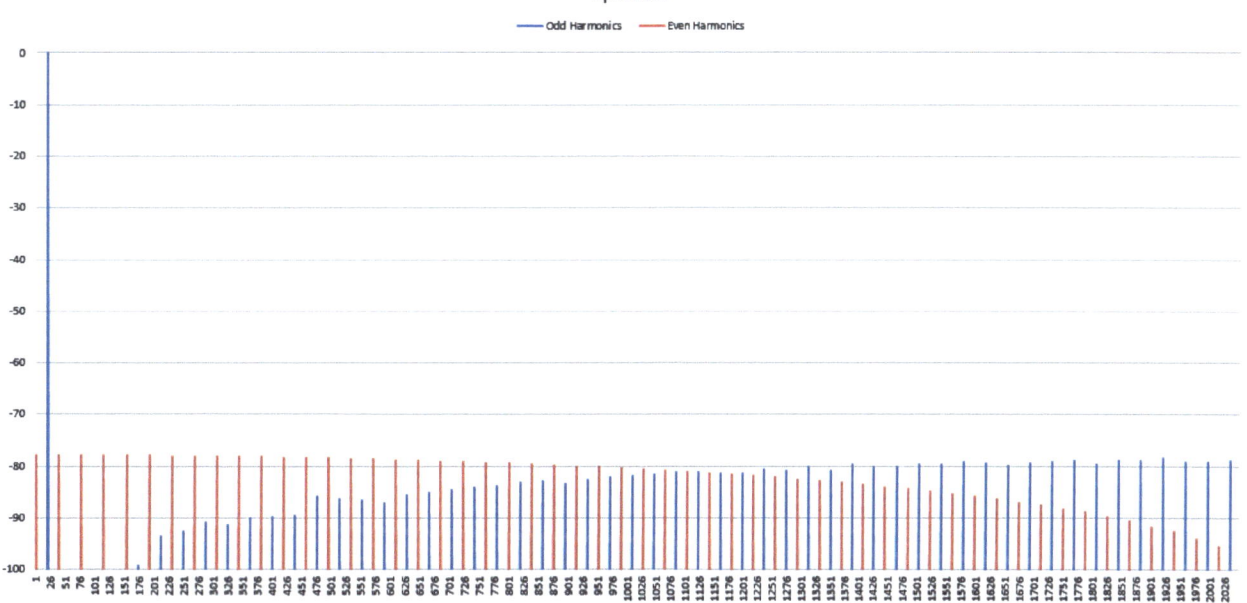

Figure 85 - 1%, 0.1uV-s TDE asymmetrical zero crossing distortion spectrum

The THD is -71.98dB for 8 harmonics and -64.07B for 50 harmonics. There is a full 8dB difference between the two values because the odd harmonics don't match the even harmonics until the 58th harmonic or so. The even THD is -64.75dB, but the odd THD is -72.47dB, a full 8dB lower because it takes so long for the odd harmonics to build up. Notice that the same level of asymmetrical distortion at the **peak** gave us a THD of -47.46dB for the first 8 harmonics and -46.34dB. Obviously, clipping is far more destructive to THD than crossover distortion.

That said, notice that the difference between the THD at 2%, 0.4uV-s TDE crossover distortion, -51.56dB, is a full 12.4dB higher (worse) than the 50 harmonic distortion value of -64.07dB for 1%, 0.1uV-s TDE. Remember with peak distortion, tests halving the distortion samples only decreased the THD by about 7dB when it should have been closer to 12dB because of the squaring factor of the area distorted. So while distortions near the zero crossover create less THD, they have a greater impact if the distortion gets better, or worse! The reason is obviously due to the fact that near the zero crossover, every "clamping" distortion that I simulate takes triangular "bites" out of the sinusoid. When the base of the bite gets longer, it also, linearly increases the opposite side due to the nearly 45 degree hypotenuse of the rising or falling slope of the sinusoid at this place in its anatomy. This means that the "bite" forms a right triangle so that linear increases in distortion cause the THD to increase by the square of the increase (12dB per doubling), unlike at the peak which changes in area nonlinearly as we saw in figure 62.

As we saw in chapter 3, at the peaks it was very easy to see the distortion. But, at the zero crossover, I have to use violet circles to point out distortions of exactly the same area, or TDE. This is the mechanism that causes distortion near the peaks to be so much more destructive to sound (or signal) quality. However, as I will point out later, at least distortion at the peaks can be avoided: simply don't get near the peaks! Audio amplifier makers have used this ploy for years. They will spec THD at some reasonable value in dB with 50% of max output—but get near the limits of the amplifier's max output; and, suddenly they are quoting THD in *percent*. That's because they have distortion anomalies at the peaks and they know it. In fact virtually everyone knows that if you turn up a radio, stereo, or other audio device to the maximum output, suddenly the sound is *bad!* What you don't want to give your customers is an amplifier that sounds bad during soft, quiet passages of music. Sure, the total energy of crossover distortion might be much smaller than the energy created during loud passages of music, but you sort of expect that. During quiet passages, your ears are more discriminating, and they will notice distortion in a very quiet moment in the music. And, besides that, crossover distortion is more destructive than many engineers believe because they never look beyond the 9th harmonic and up until that point the odd harmonics are still virtually invisible. Now you know better, and you can prove it, with a spectral display showing harmonics well beyond the 9th. What you can't see (when you aren't looking), *can* hurt you and the quality of your products.

Symmetrical Crossover Distortion

Symmetrical crossover distortion is far more likely in a Class B or Class AB amplifier, because, very often, the electronic circuitry has a symmetrical architecture. The V_{BE} drops of matched transistors that are often used in these amplifiers will almost guarantee that any distortion is going to be symmetrical. As you will see, this exacerbates the situation we saw earlier. Symmetrical distortion is going to cancel out the even harmonics; and the odd harmonics will be starting from a null near the fundamental. If you don't look beyond the traditional 9th harmonic limit for THD, you may think you have a perfect amplifier, yet your customers may disagree because they can tell there is a problem with the audio quality.

Again, let's start with the equivalent of 4% symmetrical distortion, which has 282 samples forced to zero on either side of the zero crossing, 564 samples total. The idea is to keep the area constant for this example, so that our symmetrical distortion has the same area, 1.6uV-s TDE, as the asymmetrical example at 4%, or 400 samples forced to zero above the zero crossing.

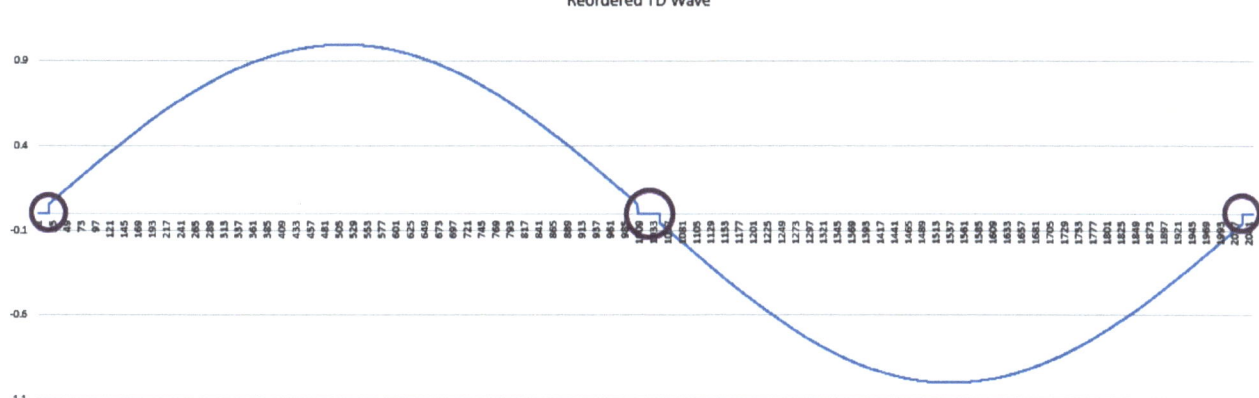

Figure 86 - 5.64%, 1.6uV-s TDE symmetrical zero crossing distortion time domain

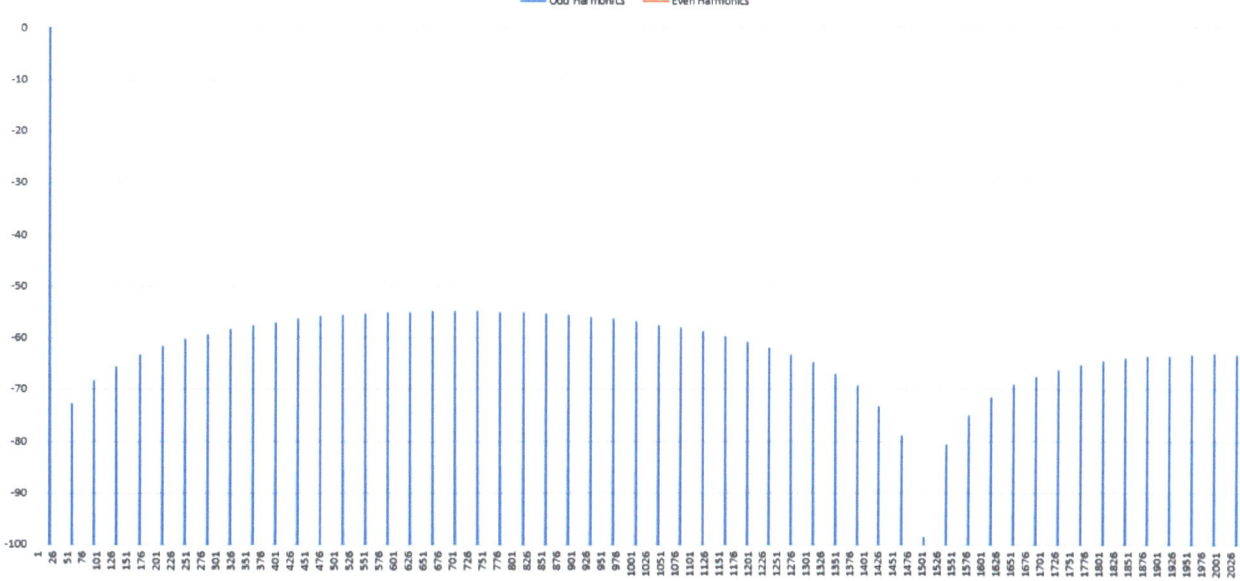

Figure 87 - 5.64%, 1.6uV-s TDE symmetrical zero crossing distortion spectrum

The THD is -60.28dB for 8 harmonics and -42.9dB for 50 harmonics. Quite a difference, and you should already know the answer as to why this is the case. There are no even harmonics at all, and the first *notch* or null in the odd harmonic signature happens right near the fundamental. So the harmonic energy builds up to a peak right around the 41st harmonic at around -54dB, which isn't even on the radar for the typical engineer, but is still audible to most people at 18KHz, given an A4 fundamental.

OK, if you test at 1KHz, the 41st harmonic won't even get through the speakers, but music does not peak at 1KHz. It peaks much lower in frequency. You can ship much higher quality amplifiers that will have a perceptible quality difference, or in the case of RF amplifiers, require less robust, and less expensive filters on the output. And, let's face it, any power that you don't have to filter out, is power that will make it to the antenna as broadcast power. In this age of battery powered devices, it's not a good idea to generate power and then simply throw it away in a filter.

Notice also, that our symmetrical crossover THD of -42.9dB, compared to the asymmetrical example which had a THD of -40.8dB, we have lost more than 2dB of harmonic energy by allowing symmetry to cancel out the **even** harmonics, even though more of the transfer function is distorted, 564 points versus 400 points in the asymmetrical example. Symmetry has advantages. Given the same total area, a symmetrical distortion creates less total harmonic energy than a distortion that is asymmetrical in either voltage or time.

Decreasing the distortion again, this time to 1.41% on each side of zero or 2.42% overall, will reduce the THD, but it will also widen the hump of **odd** harmonics, per the BHS, and make the lower frequency harmonics harder to see.

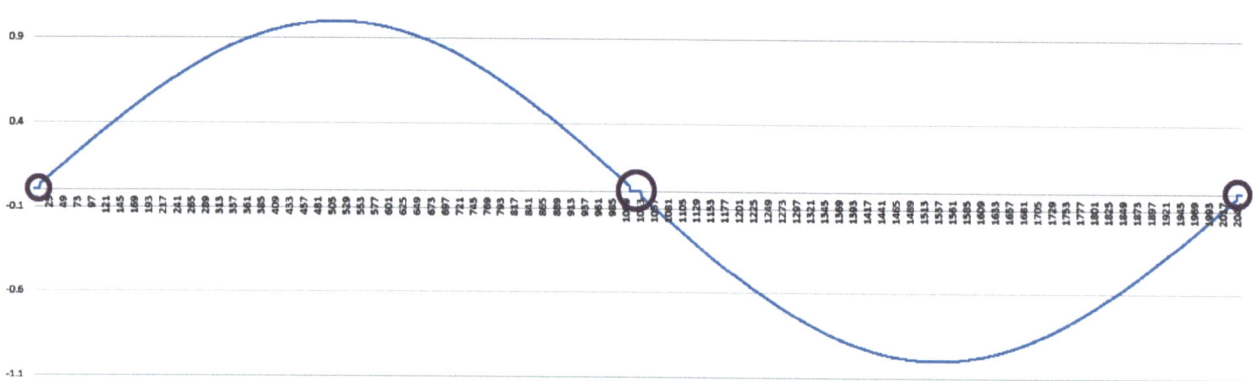

Figure 88 - 2.82%, 0.4uV-s TDE symmetrical zero crossing distortion time domain

Now for the spectrum:

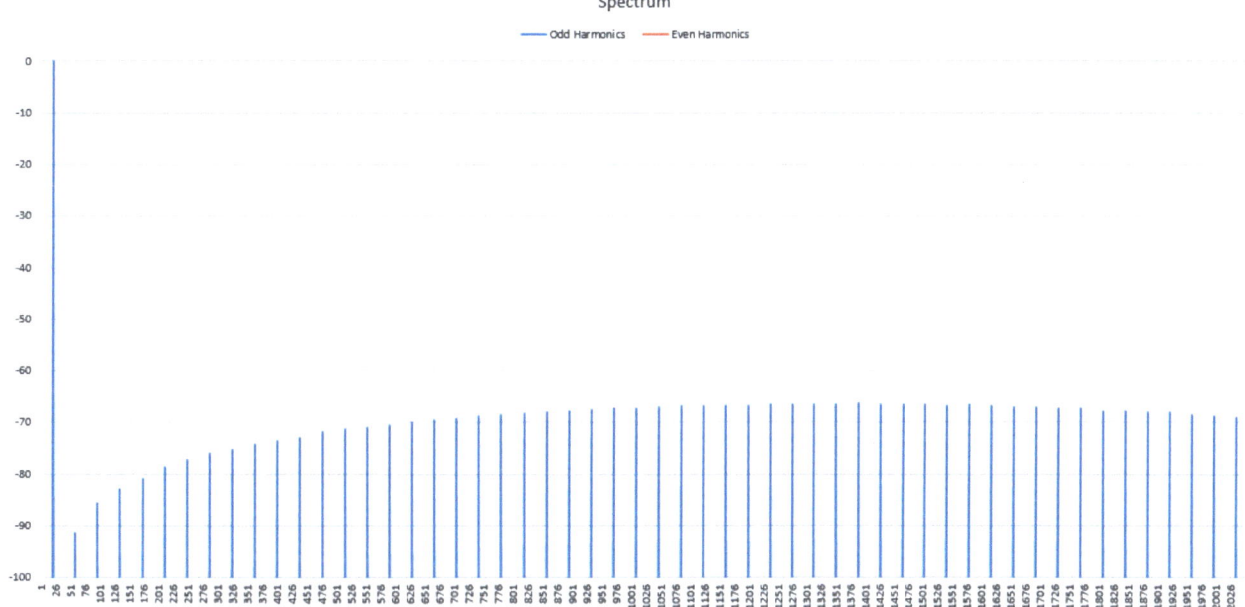

Figure 89 - 2.82%, 0.4uV-s TDE symmetrical zero crossing distortion spectrum

The THD is -77.85dB for 8 harmonics and -57.1dB for 50 harmonics. This is an impressive 21dB difference, again because the first notch, or minimum, happens right next to the fundamental instead of happening at the 13th (fig.68), 19th (fig.71), or 27th (fig.73) harmonic, as we saw in symmetrical **peak clipping** distortion back in chapter 3. This interesting behavior hides the damage done to the spectral integrity by moving the harmonic energy way above the highest place that most engineers look. The odd THD is -57.1dB, the even THD is -117.4dB, and virtually all the harmonic energy is contained in the odd harmonics; the even harmonics have all been canceled out by distortion of the opposite polarity. This assumes of course that the test wave sits at exactly zero volts offset!

You can see that this knowledge could seriously impact your engineering career, if you care about producing quality products. Your biggest challenge is going to be convincing other engineers who believe the garbage they have been taught. When you go to the management be prepared for push-back, because they will not believe you until you make this point with lots of proof. Look at all the trouble I had trying to convince my customer to believe me and not a guy with an MSEE degree. Or convince them to buy this book, hint-hint!

Now, on to the lowest level of harmonic distortion we have seen so far, 1.4% or 0.1uV-s TDE of symmetrical crossover, which is 70 samples of clamping at zero both above and below the zero crossing of the transfer function for a total of 140 samples. Notice I had to round down again, because you can't have half a sample.

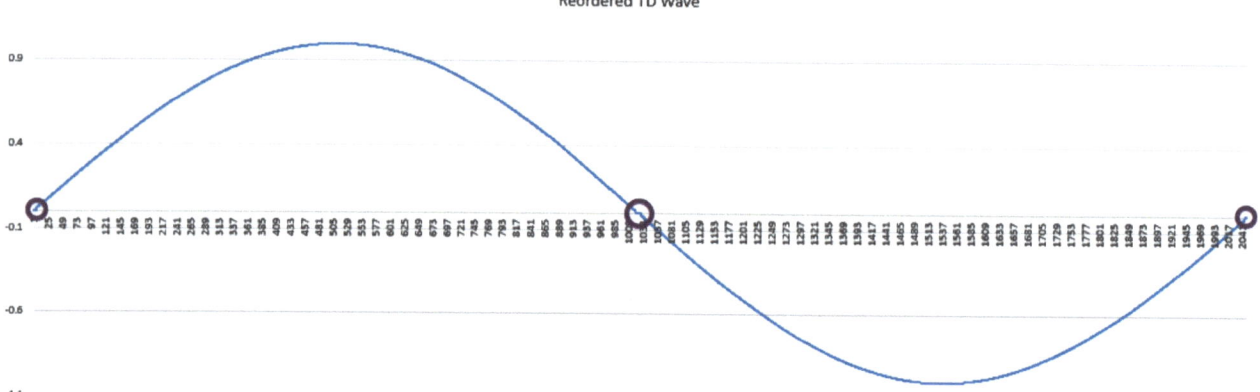

Figure 90 - 1.4%, 0.1uV-s TDE symmetrical zero crossing distortion time domain

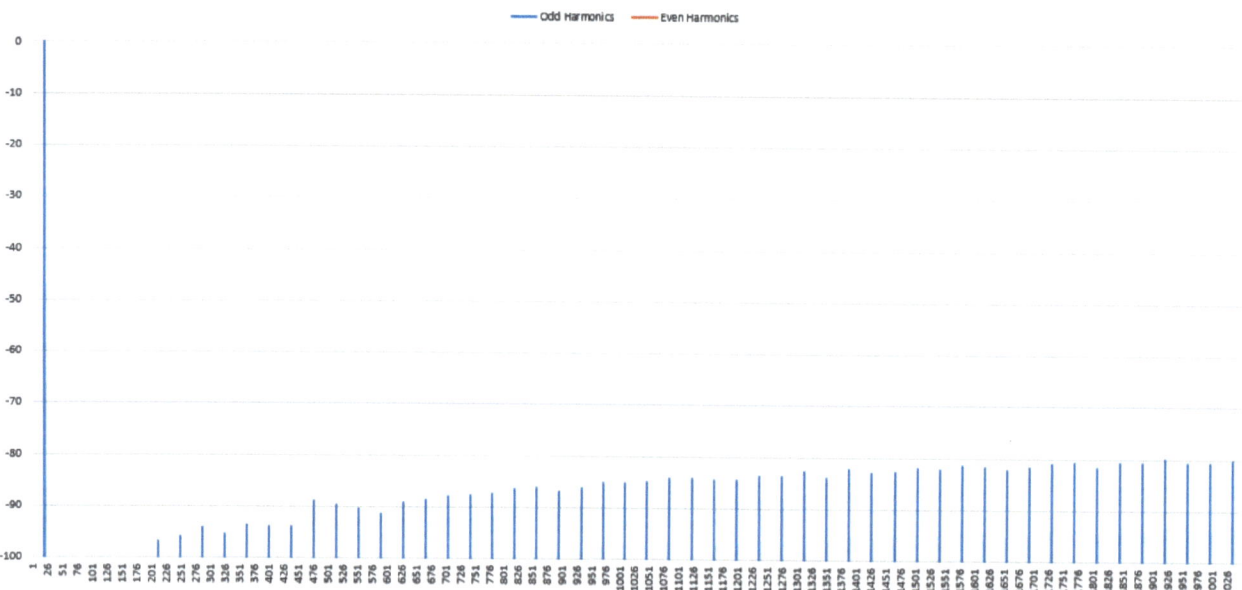

Figure 91 - 1.4%, 0.1uV-s TDE symmetrical zero crossing distortion spectrum

The THD is -100.7dB for 8 harmonics and -75.8dB for 50 harmonics. A full 25dB difference between the two numbers, because the spectrum from the fundamental to the 9th harmonic is pristine and completely free of harmonic energy. Well, sort of: I did apply an arbitrary floor of 100dB to these plots. Still, most testers can't see below -100dB so any harmonics below that level will be invisible. I will give a nod here to the Applicos[20] ATX 7006 for being the first tester to let me see to that level. Very few other testers can get down that low, and, therefore, the quality of many products suffer—to a small degree admittedly, but not imperceptibly due to this instrument noise floor.

---

[20] http://www.applicos.com

How to recognize zero crossing distortion

When you look at a spectrum analyzer or the output of an FFT on a piece of automated test equipment, you're going to wish they color coded the odd and even harmonics the way I have here. The secret to determining where the distortion comes from is watching how the odd harmonics behave. In the two most common types of distortion that I have covered here in chapters 3 and 4, the pattern is easy. We saw that even harmonics often start out high and decay down to a notch somewhere out in the spectrum, depending on the angle at which the distortion intersects the stimulating sine wave. Of course, the fact that you are seeing even harmonics tells you that the distortion is not symmetrical, or, at least, does not have enough symmetry to cancel out the even harmonics.

With peak distortion, the odd harmonics also start high and hump in a similar fashion as the even harmonics. But, if the distortion is at or near the zero crossing, the odd harmonics will be at very low amplitudes near the fundamental and will rise up in a hump further out in the spectrum according to what *angle* at which the sinusoid intersected the non-linearity in the transfer function. Remember that the severity of the distortion does not control the humpy-ness of the spectrum—only the *angle,* as predicted by the BHS.

If the distortion is between either of these extremes, the odd and even harmonics will start at some seemingly arbitrary phase relationship to the fundamental and each other and the humpy-ness is kind of hard to make out. Remember that wide slow distortions, like the INL distortion we saw in figure 15, 16 and 17 cause low frequency harmonics to show up. Short, fast distortions, like the DNL failures we saw in figures 18 and 19 cause very little low frequency harmonic distortion, but they can have quite high levels of harmonic distortion way above the fundamental and way above the 9th harmonic, where no one ever looks. Note that the DNL failure in that example happened at 45 degrees. In that case, both the odd and even harmonics start out very low, then ramp up. That was due to the angle, and you might totally miss a distortion at that point in the transfer function because you didn't bother looking far enough out.

If there is a DC offset, then that will change things because it offsets the stimulus against the transfer function. Notice that in all my experiments here, the stimulating sine wave is clamped in place relative to the transfer function. If I were to move the sine wave around relative to the transfer function, what was a symmetrical distortion would cause results similar to an asymmetrical distortion, because the sine wave is intersecting the transfer function at different parts of its *anatomy,* causing the signature to be corrupted.

This is something to keep in mind when you look at real live circuits. Remember in my video how I was able to counteract an internal 4mV mid-level offset error in the AD7671 ADC, and get the clipping distortion to be absolutely symmetrical. But now that you know more about how harmonics work, you should be able to see these anomalies as I did when I noticed that symmetrically clipping the stimulus wave on the Applicos tester still gave me even harmonics. When I saw that I knew there was an offset somewhere in the device under test, and so I twiddled the *virtual dials* in the GUI and found the offset error by pure successive approximation. Surprisingly it was my first time using that tester! Luckily I captured each screen for posterity and made a movie out of it. And, as they say, the rest is history.

Real world problems revealed

At one of my last jobs in semiconductor test, I watched my manager, a much younger guy with a BSEE degree fight an oscillating VI in current force mode[21] for months. People before him had struggled with it for at least as long, probably much longer. *Finally* he asked me for help. Thinking it might be be due to an unfortunate mix of reactive components I did the standard trick of putting different capacitors across the DUT pin, but that did nothing. Then I noticed that the VI was programmed to force 0 mAmps in the 2 Amp range. I thought that it might be a classic Class B or Class AB non-linearity right at the zero crossing. I tied in another VI and told it to pull -20mA out of the matrix pin and re-programmed the oscillating VI to force +20mA instead of 0. That did it, the previously oscillating VI quieted right down. I had solved a years old problem in two days. But why did using another VI to force current make it stop oscillating?

Because at the center of the transfer function was a tiny (~10-40mA?) non-linearity that caused the current-sense feedback network to go into a non-linear "Hunt" mode. I've seen this on an old RADAR set (SPS-10) that used a servo amplifier to drive a servomotor attached to the wiper of a large auto-transformer that acted as a line voltage regulator. It had an adjustment for the output voltage and it *also* had an adjustment for feedback speed. If you tweaked it too far one way or another it would go into a hunt mode because the servo motor couldn't keep up with the output voltage changes. In a similar way, this VI current sense circuit would try to change the output driver in a linear fashion, but the output Class AB amplifier reacted in a non-linear fashion only when programmed to zero milliAmps, right at the zero crossing, and it started hunting at the maximum speed of the negative feedback network. That's why putting capacitors on the output didn't change the frequency of the oscillation. The extra 20mA draw from the extra VI while programming the original VI to source those 20mA forced the output amplifier away from that non-linear part of its transfer function, got it into a **linear** section of its transfer function and solved the problem. A problem that not one degreed engineer had been able to solve in years, I solved it in a couple of days.

At another company I sold a tester to a customer and let another apps guy handle the app. He got stuck, his ENOB (Effective Number Of Bits) for his two 10 bit ADCs was around 4 bits, ditto for the dual DACs. He worked and worked and worked, to no avail. I flew to Taiwan, took over and found that his problem wasn't noise at all! He believed the THD function, which also returned SNR, which he converted and logged as ENOB. I found that the digital pin skew was extreme and caused some bits to get tripped later than the clock, so the sinusoid for the THD and ENOB test looked similar to figures 26 and 28, but much, much worse. What he was seeing was harmonic distortion, but after the obligatory 9th harmonic the software counts everything as noise. Harmonics are not noise, and the solution to that kind of malfunction is to fix the timing skew problem, not to add filters, or worse yet, use averaging as a solution to your ENOB problem. What does averaging do to SNR? That's right, 3.01dB improvement for each doubling of your samples. Assuming it's random noise. He found out the hard way (by losing my hard-won multimillion dollar sale) that averaging doesn't remove harmonics.

---

[21] https://youtu.be/INpsEoDJjzU

The Big Picture

Let's end this remarkable chapter with a table showing the results of our experiments thus far—and a few more I added for completeness:

Asymmetrical	Peak Clipped THD dB	Odd THD	Even THD	TDE uV-s
100 clipped at +pk	-46.34	-49.63	-49.09	0.1
200 clipped at +pk	-39.16	-42.60	-41.78	0.4
400 clipped at +pk	-32.10	-35.79	-34.52	1.6
800 clipped at +pk	-25.20	-29.38	-27.30	6.4
1600 clipped at +pk	-18.49	-23.73	-20.04	25.6
3200 clipped at +pk	-11.82	-20.63	-12.43	102.4

Symmetrical	Peak Clipped THD dB	Odd THD	Even THD	TDE uV-s
70 clipped at +pk & -pk	-47.28	-47.28	-120.19	0.1
141 clipped at +pk & -pk	-40.08	-40.08	-120.16	0.4
282 clipped at +pk & -pk	-33.10	-33.10	-120.07	1.6
564 clipped at +pk & -pk	-26.31	-26.31	-119.81	6.4
1130 clipped at +pk & -pk	-19.80	-19.80	-119.05	25.6
2264 clipped at +pk & -pk	-13.54	-13.54	-116.60	102.4

Asymmetrical	Zero Cross Clipped THD dB	Odd THD	Even THD	TDE uV-s
100 clipped at +zc	-64.08	-72.47	-64.75	0.1
200 clipped at +zc	-51.57	-54.08	-55.14	0.4
400 clipped at +zc	-40.83	-43.47	-44.25	1.6
800 clipped at +zc	-31.27	-34.11	-34.46	6.4
1600 clipped at +zc	-22.00	-24.66	-25.40	25.6
3200 clipped at +zc	-12.61	-15.05	-16.28	102.4

Symmetrical	Zero Cross Clipped THD dB	Odd THD	Even THD	TDE uV-s
70 clipped at +zc & -zc	-75.82	-75.82	-117.37	0.1
141 clipped at +zc & -zc	-57.09	-57.09	-117.37	0.4
284 clipped at +zc & -zc	-42.91	-42.91	-117.37	1.6
564 clipped at +zc & -zc	-33.09	-33.09	-117.37	6.4
1130 clipped at +zc & -zc	-23.35	-23.35	-117.33	25.6
2264 clipped at +zc & -zc	-13.72	-13.72	-117.00	102.4

This chart clearly shows that distortion near the edges of the transfer function is significantly more serious than crossover distortion. And yet, crossover distortion is not as inconsequential as most engineers believe. The problem as we have seen, is that most engineers simply *assume* (which makes an *ass* out of *u* and *me* as they say) that harmonics harmlessly roll off as though there was some inherent low pass filter in the formula that describes harmonics. This could not be further from the truth, and now you are among the few who understand this fallacy.

It also shows that *area* is what matters when it comes to distortion. I color coded the chart so that all rows of the same color have the same area. Notice that in Peak Clipped waves with 0.1uV-s of area distorted, the 50 harmonic THD for asymmetrical clipping at -46.34dB is almost identical to the symmetrical clipping example, -47.28dB, despite the fact that the symmetrical clipping example involved 40 more points of distortion in the transfer function. Remember, that was 70 points on the top and bottom for a total of 140 samples, versus 100 samples in the asymmetrical clipping example. But, if you go down to the Zero Cross Clipped section, you can see that there is a huge difference between the THD for symmetrical vs asymmetrical at the lower levels of distortion. Notice also, how, as distortion increases, the difference between Peak distortion and Zero Crossing distortion almost disappears. At 6.4uV-s, which represents over 10% of the transfer function compromised, the numbers for Peak and Zero Crossing THD are very close.

And finally, look at those **even** THD numbers for symmetrical distortion, both Peak Clipping and Zero Cross Clipping. Symmetry cancels out all **even** harmonics reducing the THD, proving again that my video rebuke of Audio Precision's book was justified.

I would submit that it is a waste of time to even look at **even** harmonics when measuring THD. How much energy there is in the **even** harmonics is simply a measure of asymmetry, nothing more. If you want to really know how bad the distortion is, measure THD using a bunch of **odd** harmonics and be done with it.

But really I think THD is worthless anyway, what's the point? It's a **lie**!

Chapter 5
The Lie of THD

Read any analog component spec and you'll find a THD number. Buried in the plethora of notes and caveats are the conditions under which THD was tested. Waveform frequency, number of harmonics counted, RMS amplitude of the test signal, DC offset, load on the output, impedance of the input, etc, etc. At least it's more honest than most automobile MPG numbers, where the conditions are assumed or implied by some incomprehensible set of government mandated conditions. Why is this the case? Why does the manufacturer have to specify every single parameter when testing THD?

Remember how even just a slight alteration in the angle of where the sinusoid hits an equivalent area distortion in the transfer function changes the THD? We saw this in Chapter 4 with this table where I moved a 0.1uV-s non-linearity across the face of our test sinusoid and watched the 50 harmonic THD change dramatically.

Angle	Area	THD	Area Delta/ THD Delta
78.5	855.8nV-s	-46.24dB	NA
73.7	280nV-s	-52.72dB	0.33 / 0.47
70.1	214nV-s	-54.42dB	0.77 / 0.82
66.9	179nV-s	-55.7dB	0.84 / 0.87
64.2	153nV-s	-56.95dB	0.86 / 0.87
61.6	134nV-s	-58.04dB	0.87 / 0.88
59.3	130nV-s	-58.2dB	0.97 / 0.98
57.1	124nV-s	-58.52dB	0.96 / 0.96
55.1	125nV-s	-58.48dB	1.0 / 1.0
53.1	115nV-s	-59.06dB	0.92 / 0.94

If I try to duplicate THD using a different instrument than what was used at characterization, one tiny variable set to the wrong value and it's all over. If I am off just a tiny amount in the amplitude of my "activation" sinusoid, I will get a different answer, and that can make people crazy trying to understand what went wrong with correlation. It can also give people the wrong idea about my quality, maybe for nefarious purposes.

Let's say that I have a company called "Classy-A Microelectronics," where I make amplifiers based on a Class A topology which I consider superior to Class B or Class AB amplifiers. I know that my amplifiers will have issues at the edges of the transfer function, just like our 5.64% symmetrical peak clipping transfer function; but I have to try to beat my competitor's "B-Boyz Amps" Class B amplifiers that have a similar 5.64% symmetrical transfer function glitch—but at the crossover. Can I beat their THD numbers?

The answer is **no**, unless I *cheat!* Remember that with a 5.64%, 1.6uV-s TDE peak distortion in the transfer function I get a THD number of -33dB; but, because of the way harmonics work, an equivalent crossover distortion of 5.64%, or 1.6uV-s TDE gives me a 8 harmonic THD value of -60dB. B-Boyz is going to take all my customers away because, even though their distortion happens right in the middle of the transfer function, it has less impact on the THD than my parts, because of the greater area of the waveform at the peaks. Now, I could try to explain to my potential customers that they could easily avoid my distortion by just using a slightly lower amplitude wave, a wave that will not even hit those distortions. This is similar to the advantage a motorcycle has over a car when it comes upon a speed bump. Speed bumps have gaps to prevent pooling when it rains, and a motorcycle can zip down a street with speed bumps at 90MPH by driving through the gaps, whereas a car might be able to avoid putting one pair of tires on the speed bump, but the other side of the car will be fully impacted. The driver might enjoy splitting the speed bumps by driving his left wheels (in the US) through the gap, but his wife will not be amused when the right side of the car gets hammered by the speed bumps on her side at 90MPH.

So, at Classy-A Microelectronics, I specify that THD is measured using a waveform with an amplitude of 90% of full scale, avoiding the peak distortion with a few percent to spare for offset issues, and I get away with a THD of -100dB. B-Boyz has to admit that their Class B amplifiers are saddled with a THD of -60dB when tested with a sine wave at 100% of full scale. When their potential customers ask why they don't follow the lead of Classy-A and use a test signal at 90% of full scale they have to admit that reducing the amplitude of the fundamental will exacerbate the problem. Classy-A's THD numbers get dramatically better when they reduce the amplitude of their test tone; but B-Boyz amplifiers can't easily avoid the zero crossover. To make matters worse, because THD is measured as a ratio of the total harmonic power divided by the total fundamental power, reducing the power of the fundamental will not help; it will actually make the THD numbers worse! If B-Boyz could reduce the amplitude of the signal to less than half the full range and offset it above the zero crossing distortion, they *might* be able to get to a THD of -100dB; but, someone is likely to pick up on that fact, and now their SNR value will be at least 6dB worse, because of the lower signal amplitude relative to minimized, but irreducible, noise level.

I am almost certain that this is how we ended up with virtually all amplifiers based on Class B or Class AB topology. Sure, the THD numbers are better for Class B over Class A amplifiers. The only problem is that the THD advantage that Class B amplifiers have over Class A amplifiers is a **lie**! Total Harmonic Distortion has *some* validity, but not much. It's too easy to cheat, too easy to adopt inferior topologies that severely distort the waveform, especially quiet waveforms. Think about a very quiet passage of music going through a B-Boyz amplifier. It will get beaten to death by the crossover distortion, while the same quiet passage will sound absolutely beautiful coming out of a Classy-A amplifier. And, yet, if THD is measured the same way, the Classy-A amplifier spec sheet suffers severely, while the B-Boyz amplifier looks great. On paper, that is.

Does this mean I am advocating for Class D amplifiers? Are you kidding? Remember when we looked at the spectral signature of square/rectangular/pulse waves, we saw that the spectra for those waves were very ugly. They spew harmonics all over the spectrum, and, as the PWM mechanism moves those very fast edges back and forth in an effort to simulate sinusoids, the harmonics bounce all over the place. Take a look at the THD spec for a typical Class D amplifier, 1% (-40dB) at 50% volume, 10% (-20dB) at 100% volume. Anytime you see a THD spec in percent you know the manufacturer doesn't care about quality. Class D amplifiers might really be linear, but using a speaker as a parasitic low pass filter and hoping that it takes away all the harmonics generated by PWM is being somewhat optimistic. I prefer Class A.

Discarding THD

I think I have proven to you that THD is just a bad way to go, but, what is the alternative? Do we have to invent a new standard measuring stick to replace the rubber bands (THD) we have been using up till now?

We have all the tools we need to differentiate a good amplifier from a bad amplifier. We already use them in a slightly different form to test DACs and ADCs. Everything we need to know is contained in one parameter that we capture any time we test a DAC or ADC. We just have to get used to applying the same test to other analog components.

If we perform a standard ramp or sine wave based linearity test, we can accumulate all the non-linearities into an array. However, rather than just looking for the worst case non-linearity, which tells us nothing about *area*, we sum up the *absolute* linearity errors over the whole transfer function, and record the value as Total Distortion Energy (TDE) in milliVolt-seconds, or mV-s.

Remember how we learned that it was *area* that causes harmonic distortion, and, in electronics, area is energy. The integral (sum) of all non-linearities tells us how much energy will be created by the distortions inherent in the device (amplifier, DAC, ADC, multiplexer, etc) and how much it will negatively impact *any* waveform we pass through it. Despite its name, Integral Non-Linearity (INL) is not the integral of anything and tells us very little about how much harmonic energy will be created by transfer function non-linearities.

As I mentioned in chapter 3, by scaling TDE in milliVolt-seconds, we imply a 1KHz reference tone, whether we test with a 1KHz tone or not. We can test with a ramp, a triangle wave, or use a 1KHz sine wave to perform a sinusoidal histogram just as we would do with an ADC. This would satisfy the naysayers who insist on a dynamic test strategy. The sum of all the absolute non-linearities is measured and scaled in milliVolt-seconds, and therefore implies that, if a 1KHz sine wave were applied across the entire transfer function, it would insert **this** amount of distortion energy into the signal which would become a proxy for the very flawed measurement we now use, THD.

THD can't test the very edges of the transfer function, which is the dodge used by our imaginary Classy-A Microelectronics, because you can't really test the very edges of any amplifier, lest you accidentally clip due to an inherent offset in the circuit—as you should have seen in my video. An amplifier that has a 4mV mid-level offset (as in my video) requires that in any THD test, you reduce the amplitude by 8mV peak-to-peak because if you don't, you will clip on one side, which will make the THD worse than if you over-drive the circuit to take advantage of the fact that symmetrical distortion removes the even harmonics and thus makes the THD numbers look better. Using a linearity test to measure TDE by going all the way to the edges of the transfer function will tell the truth. It also negates the advantage of a transfer function that has symmetrical distortions. Symmetry is a relative concept.

Remember that we don't always send pure, zero offset sine waves through amplifiers. A symmetrical crossover distortion is symmetrical only for signals that have a zero volt offset. Remember in my video, it only took a 4mV offset to cancel the advantage of symmetry. What happens when a violin plays in concert with a cello? As the low frequency cello signal oscillates up and down, the higher frequency violin tone that is riding on the lower tone crosses every place on the transfer function; so, while the violin tone alone *might* benefit from the symmetry of a distortion symmetrical about zero, it cannot benefit from that symmetry when it's riding on a lower frequency tone that takes it above and below the fixed symmetrical non-linearity in the transfer function.

Apples and Oranges

Remember this chart I created that compared four kinds of distortion? I have asymmetrical peak clipping, asymmetrical zero crossing clipped at zero, symmetrical peak clipping, and symmetrical zero crossing clipped at zero:

Peak Clipped					Peak Clipped				
Asymmetrical	THD dB	Odd THD	Even THD	TDE uV-s	Symmetrical	THD dB	Odd THD	Even THD	TDE uV-s
100 clipped at +pk	-46.34	-49.63	-49.09	0.1	70 clipped at +pk & -pk	-47.28	-47.28	-120.19	0.1
200 clipped at +pk	-39.16	-42.60	-41.78	0.4	141 clipped at +pk & -pk	-40.08	-40.08	-120.16	0.4
400 clipped at +pk	-32.10	-35.79	-34.52	1.6	282 clipped at +pk & -pk	-33.10	-33.10	-120.07	1.6
800 clipped at +pk	-25.20	-29.38	-27.30	6.4	564 clipped at +pk & -pk	-26.31	-26.31	-119.81	6.4
1600 clipped at +pk	-18.49	-23.73	-20.04	25.6	1130 clipped at +pk & -pk	-19.80	-19.80	-119.05	25.6
3200 clipped at +pk	-11.82	-20.63	-12.43	102.4	2264 clipped at +pk & -pk	-13.54	-13.54	-116.60	102.4

Zero Cross Clipped					Zero Cross Clipped				
Asymmetrical	THD dB	Odd THD	Even THD	TDE uV-s	Symmetrical	THD dB	Odd THD	Even THD	TDE uV-s
100 clipped at +zc	-64.08	-72.47	-64.75	0.1	70 clipped at +zc & -zc	-75.82	-75.82	-117.37	0.1
200 clipped at +zc	-51.57	-54.08	-55.14	0.4	141 clipped at +zc & -zc	-57.09	-57.09	-117.37	0.4
400 clipped at +zc	-40.83	-43.47	-44.25	1.6	284 clipped at +zc & -zc	-42.91	-42.91	-117.37	1.6
800 clipped at +zc	-31.27	-34.11	-34.46	6.4	564 clipped at +zc & -zc	-33.09	-33.09	-117.37	6.4
1600 clipped at +zc	-22.00	-24.66	-25.40	25.6	1130 clipped at +zc & -zc	-23.35	-23.35	-117.33	25.6
3200 clipped at +zc	-12.61	-15.05	-16.28	102.4	2264 clipped at +zc & -zc	-13.72	-13.72	-117.00	102.4

Remember the extra advantage that crossover distortion gave us? Because the sine waves are moving fast near the center, Crossover Distortion created most of its odd harmonics higher up in the spectrum, fooling us into believing that Class B amplifiers are better than Class A. Even non-symmetrical distortion near the zero crossing benefits from this by suppressing the odd harmonics near the low end of the test spectrum. By testing only 8 harmonics, or 7 harmonics, or, even worse (as some companies do), 5 harmonics, symmetrical crossover distortion can be almost entirely missed—as you saw in chapter 4. Remember that as the distortion gets closer to the 180/360 degree crossing, the harmonics zoom off to the right and any THD test will miss much of the energy you are generating, and now you are knowingly, or unknowingly counting on parasitic low pass filters to rid you of it. In fact, that is a common dodge used in THD testing, ignoring troublesome harmonics; and yet the harmonics are there, further up in the spectrum—but nobody knows about them (or maybe they do and they are just hoping you are not going to look there). The really bad news is that if an SNR test is done along with the THD test, those extra harmonics are counted as noise, and that certainly does not help the SNR spec!

THD really has no validity. But TDE will give you the Total Distortion Energy no matter where the distortion is in the transfer function. TDE will not allow cheating, it tests every point in the transfer function, far more than any sine wave can (remember figure 74!) and it uses a linear function, a ramp, to do it.

Advantages of understanding Distortion

In a mobile RF amplifier (cell phones, hotspots, etc) power is everything, more power to the antenna and less power from the battery are the biggest concerns. So when you have a Class B amplifier, or worse yet, a Class C, you are spewing tons of energy into a filter that you had to design to mash all that unwanted garbage down. Why generate energy when you are just going to throw it away? Even a nicely biased Class AB amp will have *some* crossover distortion, and that will create energy far out into the spectrum. Where is it going? Do you even realize it's there? Have you been assuming that it will never appear since it wasn't causing problems in the second, third or fourth harmonic? Remember that if the two transistors are well matched (symmetrical) your **odd** harmonics will start near -100dB, then ramp up to some respectable level of power way out in the spectrum. "Oh, well, it doesn't matter because my antenna and other parasitic elements will attenuate it down to nothing." Yeah, well your battery is being drained for no good reason just to generate harmonics you don't even want coming out of your application. Despite the fact that **even** harmonics don't appear, they still get generated, it's just that we can't see them because they cancel each other out. Is that a good use of power? I don't think so. And if you don't generate *any* harmonics, you don't have to filter them. Smaller, or possibly **no** filters are possible once you know what your amplifier is spewing out into the universe.

In FMK (coherent Frequency Modulation Keying) we switch the frequency of the carrier at the zero crossing. Most audio multiplexers switch at the next zero crossing. Is that a good idea? *Are you sure?* Maybe you need to learn about The Bullard Effect[22] before you answer that question.

In audio applications we can **finally** understand why some people prefer tubes over transistors, Bipolars over MOSFETS, vinyl to digital, etc. Then we can stop bickering about it by offering unproven hypotheses, conjecture or hearsay as evidence, we *will finally know!*

Remember the story I told about that Class AB VI that tended to oscillate when programmed to zero milliAmps in the two Amp range? This was on a military IC test platform. What if instead of being designed into part of the test platform it had been built into the actual military ordinance? Now you have a missile steering fin being driven by an unstable feedback network and who knows what it's going to do when launched? If it's carrying a nuclear warhead do you think the stakes are high enough that we should understand what's going on in these ubiquitous and **heretofore untested** Class AB amplifiers?

---

[22] https://youtu.be/wrl_Es4CZvg

Changing the world

We can make the world better, but only if people understand how harmonics work. You are now among the few who have a good idea how harmonics work, how symmetrical distortion can cancel **even** harmonics, making your THD numbers look better, and how crossover distortion and other short duration, fast edged distortions look better compared to peak distortion if you only look at the first few harmonics. Not everyone understands, and that is *their* deficiency, not yours.

When I was forced to dive into the cause of harmonics, I had no idea where it would lead. I was simply trying to save my reputation. I take it rather personally, partly because I don't have a degree which means that I don't have a piece of paper that someone signed saying I know what *they* know. But that has turned out to be a huge advantage. In my mind, the things your professors told you are all in question, and you have used them as assumptions throughout your career without understanding why you believed them. If you didn't believe them, you failed the test and didn't get your diploma. We know that Mr. Tibbetts got his diploma. Who was right? *Can* adding a DC offset change the harmonic signature of a waveform applied against a flawed transfer function? I think you know the answer to that question, now.

As I did in my Applicos video, let me quote Samuel Clemens, aka Mark Twain, on knowledge:

"It ain't what you don't know that gets you into trouble. It's what you know for sure that just ain't so."

Once I figured all this out I was stunned at how ridiculous a measurement THD was. Yet, for years I had chastised people who used INL and DNL to measure DAC and ADC quality. I had a superior attitude because I had written my own FFT, I had written my own THD functions. I had to: nobody else could track aliasing harmonics, which got lost beyond the Nyquist frequency and got counted as noise in the SNR calculations. Those guys who used INL and DNL were just chumps. Later, I realized that THD, even if it was hard to do, was just wrong.

Once I understood what was at stake, I went to Applicos and told them that together we could **change the world** because I have always wanted to be one of the *Crazy Ones*[23]. Failing there I went to Maxim and asked them to help me change the world. They were a little more receptive, but still, I didn't make my point. I had to write this book to make my case. You must judge if I was successful.

So here I am, asking you to help me change the world. Analog is almost passé, microphones are now digital. Amplifiers are going Class D (God help us), which is purely digital. We should fix this before we go so far away from analog that no one will understand where the errors in the reasoning are, where the bad assumptions are, before we run into a problem that results from something that we *"know for sure that just ain't so."*

---

[23] https://youtu.be/8rwsuXHA7RA

About the author

Dan P. Bullard has worked in electronics since 1976. He was educated at Delta Junior College in Stockton California and US Navy Electronics Technician school in Great Lakes Illinois. Dan was always at the top of his class and proved his deep understanding of electronics over and over again at US Navy installations at San Nicolas Island California, Harold E. Holt Communications Station in Exmouth Western Australia, and Mare Island Cryptographic Technical School in Vallejo California as well as in the private sector at Technical Training Center, GenRad, Credence Systems Corporation, Schlumberger, Intel, Nextest Systems Corporation, Maxim Integrated Products, Tektronix and Applicos.

Dan is an expert at Digital Signal Processing, holds a patent in Smart Card/RFID testing and has been published in various electronic trade magazines as well as Houseboat Magazine. Dan's travels are documented in The Reluctant Road Warrior, available on Amazon.com. Dan has a YouTube channel[24] that specializes in electronic topics and owns the domain www.danbullard.com where you can find many articles, tutorials, animations, and videos on electronic topics.

Dan on the Columbia River near his home in Vancouver Washington

---

[24] https://www.youtube.com/channel/UCF1enwmIcIeupPESjk5QqyA or http://tinyurl.com/nlx9stx

www.ingramcontent.com/pod-product-compliance
Lightning Source LLC
Chambersburg PA
CBHW050730180526
45159CB00003B/1186